W9-AAX-867

TARGET: PRIME TIME

COMMUNICATION AND SOCIETY
Edited by Gerge Gerbner and Marsha Siefert

IMAGE ETHICS
The Moral Rights of Subjects in Photographs, Film, and Television
Edited by Larry Gross, John Stuart Katz, and Jay Ruby

CENSORSHIP
The Knot That Binds Power and Knowledge
By Sue Curry Jansen

SPLIT SIGNALS
Television and Politics in the Soviet Union
By Ellen Mickiewicz

TARGET: PRIME TIME
Advocacy Groups and the Struggle over Entertainment Television
By Kathryn C. Montgomery

TELEVISION AND AMERICA'S CHILDREN
A Crisis of Neglect
By Edward L. Palmer

PLAYING DOCTOR
Television, Storytelling, and Medical Power
By Joseph Turow

TARGET: PRIME TIME

Advocacy Groups and the Struggle over Entertainment Television

Kathryn C. Montgomery

New York Oxford
OXFORD UNIVERSITY PRESS
1989

Oxford University Press

Oxford New York Toronto
Delhi Bombay Calcutta Madras Karachi
Petaling Jaya Singapore Hong Kong Tokyo
Nairobi Dar es Salaam Cape Town
Melbourne Auckland

and associated companies in
Berlin Ibadan

Published by Oxford University Press, Inc.,
200 Madison Avenue, New York, New York 10016

Library of Congress Cataloging-in-Publication Data
Montgomery, Kathryn.
Target : prime time : advocacy groups and the struggle over entertainment television /
Kathryn C. Montgomery.
p. cm. Bibliography: p.
Includes index.
ISBN 0-19-504964-0
1. Minorities in television. I. Title. II. Title: Advocacy groups and the struggle over entertainment television.
PN1992.8.M54M66 1989
305.8'0973—dc 19 88-23287 CIP

2 4 6 8 9 7 5 3 1

Printed in the United States of America
on acid-free paper

Preface

In 1977, prime-time television was under attack on several fronts: the PTA had just put the networks "on probation" for excessive violence; the U.S. Civil Rights Commission had accused the TV industry of discrimination against women and minorities; and church groups were trying to keep the provocative new ABC comedy, *Soap,* from reaching the airwaves. While these events were making headlines, I was immersed in doctoral coursework at UCLA and teaching part-time at California State University, Los Angeles.

As a student and teacher of the mass media, I was struck by the fact that, with all the discussion and debate about "pressure groups," there was very little understanding of their impact on television. I was also aware of a marked dissonance between the official statements network executives made to the press and the first-hand accounts I was getting from friends in the television business. And I could see that many people outside the industry—including a number of the advocacy groups trying to influence television—were confused about the way the networks operated, often acting according to conventional wisdom which was usually inaccurate or incomplete.

I was curious about several things: Were advocacy groups having any significant impact on entertainment programming, or were they just a minor aggravation to the networks? Were these groups—as was often charged—responsible for making TV programming bland, or were they having a more positive influence? How was the powerful institution of network television coping with pressures from outside?

I began to pursue these questions as a possible doctoral thesis. My professors were understandably skeptical about such a project, warning me that few people in the industry would agree to talk to me on the record about this issue. And some of my first interviews at the networks seemed to bear this out. "For all I know," charged one executive as he eyed me suspiciously, "you could be writing a blueprint for pressure groups!" Eventually I decided to focus my dissertation research on a case study of the gay activists, reported to be one of the most successful advocacy groups. I found gay organizations extremely cooperative, not only consenting to lengthy interviews but also making their files (which included extensive correspondence with the networks) available to me. This information, along with interviews of network executives, producers, and writers, formed the basis of my dissertation, which was completed in 1979. (Part of that research has been incorporated into Chapter 5 of this book.)

At the end of my study I was even more intrigued. I knew that I had been able to see only a small part of the entire picture. The experience had given me a fresh perspective on the subject, opening up new doors and raising more questions. It had also sharpened my research skills and equipped me with a basic understanding of how the system worked.

During the next few years I avidly followed developments between advocacy groups and network television. Although I didn't have the time to do more formal research, teaching in Los Angeles made it possible to keep track of ongoing events at close hand. I frequently had TV industry representatives and advocacy group leaders lecture to my classes. And I was present at several critical events.

In 1980, I took my students to hear Federal Communications Commission chairman Charles Ferris address a group of blacks in the industry. It was during this event that the controversy over NBC's *Beulah Land* suddenly burst into the public spotlight. As the battle escalated, I followed it closely.

The next year, I participated in a dramatic three-day conference on "The Proliferation of Pressure Groups in Prime Time." At this meeting—open to the press but closed to advocacy groups—network executives, producers, writers, advertisers, journalists, and scholars hotly debated the issues raised by the recent wave of pressures on the industry.

In 1982, I attended a follow-up meeting in Aspen, Colorado. This time, a handpicked set of advocacy group leaders had been invited to sit down with network representatives in order to hammer out "rules of the game" for dealing with network television. The press was barred from attending. What became clear to me was that there were already in place unwritten industry rules governing the relationship between advocacy groups and the networks. These "rules of the game" had never been made public.

By this time I had become so engrossed in the subject that I knew I had to write a book. My appointment to the UCLA faculty gave me the time and institutional support I needed to accomplish such a major project.

Since virtually no research had been done in this area, I had few models to follow for my work. Fortunately, there were some studies on decision making in television that I was able to draw upon for background and guidance. During the early stages of my doctoral research, for example, I found Edward Jay Epstein's pioneering 1974 book, *News from Nowhere,* particularly insightful at documenting the role played by organizational structures and policies in shaping network news stories and broadcasts. The work of Muriel Cantor, especially her study of Hollywood TV producers, was also an important influence. George Gerbner's study during the 1950s of the operations of the network censorship departments and Robert Pekurny's 1977 doctoral dissertation on the structure and operation of NBC's Broadcast Standards department both gave me useful background material.

As I continued my own investigation, I was encouraged by the work of a small, but growing, group of media scholars who were also examining the internal processes of entertainment television. Geoffrey Cowan's book, *See No Evil,* gave a lively and informative behind-the-scenes account of the pressures over TV violence. Joseph Turow's articles and his book, *Media Industries,* provided a comprehensive description and analysis of institutional structures and processes. Horace Newcomb and

Robert S. Alley's collection of interviews with prominent television producers *(The Producer's Medium)* provided a context for better understanding the production community. And finally, Todd Gitlin's 1983 study, *Inside Prime Time,* was particularly valuable for its "thick description" of the televison industry subculture.

I conducted my research on two levels. The first was designed to establish the broad contours of the relationship between advocacy groups and network televison. I talked to as many group leaders and industry representatives as possible, and surveyed a wide range of primary and secondary published material. My second level of research was a series of detailed case studies examining the most significant groups, protests, and pressure campaigns.

Much of the primary data in this book comes from my interviews with advocacy group leaders, network executives, producers and writers. Given the highly charged nature of the subject matter, some people declined to be interviewed. Others could not be found. But a surprisingly large number of people did agree to talk to me. While interviews are an effective means of understanding peoples' attitudes and perceptions, they are not always sufficient as research sources, because of selective memory and intentional or unintentional distortions. To compensate for this limitation—and for the unavailability of some participants—I corroborated many of the facts from interviews with accounts from others involved in particular incidents. I also used information from newspapers and trade publications, as well as primary documents such as scripts and correspondence.

As the title of the book suggests, this study is restricted to efforts at influencing network *prime-time* programs. For that reason, groups involved with other areas of television programming have been mentioned only in passing. I also touch only briefly on those aspects of early TV history that others have written about at length. (For example, I do not spend much time on the anti-communist blacklist in television.)

A word about citations. At the suggestion of my editors, in

an attempt to make the book easier to read, I have collapsed the references and explanatory notes for each paragraph into a single entry. These appear in the Notes section at the end of the book.

The story that unfolds within these pages is a continuing one. As I write these words, new developments are taking place, adding twists and turns to the ongoing narrative. There will be many further changes over time, some of which I've tried to anticipate in the last chapter. These will be strongly influenced by the relationships, mechanisms, and processes that I have documented. It is my hope that this book will prove valuable—to the general public, to students of the media, and to policy-makers—for its insights into an area that has remained murky and misunderstood for too long.

Los Angeles, California
August 1988

Kathryn C. Montgomery

Acknowledgments

Many people—within the television industry, the advocacy group community, the government, and the press—were generous in talking to me about this research. Some are cited within these pages, others are not. But all of them contributed in some way to the final product. I offer my sincere thanks and appreciation to: Ray Andrade, Warren Ashley, Josh Baran, Eleanor Blumenburg, Gina Blumenfeld, Kathy Bonk, Emma Bowen, Lydia Bragger, Tyrone Braswell, Warren Breed, Barbara Brogliatti, James Brown, Mary Ann Brussat, Robert Butler, Virginia Carter, Peggy Charren, Aaron Cohen, Tony Cortez, David Cuthbert, Christopher Davidson, James DeFoe, Newton Deiter, Dara Thorpe Demings, Henry Der, Daniel Donehey, Jean Dye, Tilden Edelstein, Sam Elfort, Dwight Ellis, Jan Engsberg, Norman Fleishman, William Fore, Grace Foster, Francesca Friday, Bill Froug, Kathy Garmezy, David Gerber, Richard Gilbert, Richard Gitter, Ron Gold, Maurice Goodman, Judy Greening, Milton Gross, Herb Gunther, Felix Gutierrez, Tari Susan Hartman, Sumi Haru, Charlie Hauck, Helen Hernandez, Bettye Hoffman, Patty Hoffman, Julie Hoover, Wiliam Hutton, Maggie Inge, Reed Irvine, Nicholas Johnson, Robert Kalaski, Richard T. Kaplar, Marcy Kelly, Tom Kersey, James Komack, Norman Lear, David Levy, Loretta Lotman, Jim McGinn, John McGreevey, Brian Malloy, Leonard Matlovich, Stan Margulies, Pluria Marshall, Stephen Menick, Tom Moore, Barbara Moran, Susan Newman, Frank Orme, Joe Partansky, Jack Petry, David Poindexter, Robert Price, Jay Rodriguez, Meta Rosenberg, Barney Rosenzweig, Guadalupe Saavedra, Alfred Schneider, Sam Simon, Eva Skinner, Sally Steenland, Jerilyn Stapleton, Larry Stewart, Horst Stipp, Suzanne Stutman, Herminio Traviesas, Chris Uszler, Nicholas Van Dyck, Ginny Vida, Joe Waz, Donald Wear, Howard Woolley, and Collette Wood.

I could not have completed this project without the support of the University of California, which awarded me research and travel grants and provided clerical assistance. I am also grateful to the UCLA Theater Arts library and especially to the Terrence O'Flaherty Television Collection.

I am thankful for materials and research assistance provided by: the Television Information Office, the Broadcast Pioneers Library, the National Association of Broadcasters, the American Broadcasting Company, the Columbia Broadcasting System, and the National Broadcasting Company.

Among my colleagues were a number of people who gave me advice, feedback, and direction. I am particularly grateful to: Muriel Cantor, Ruth Schwartz, Charles Firestone, Paul Rosenthal, Janet Wasko, Dan Schiller, Erik Barnouw, Joseph Turow, Geoffrey Cowan, William Froug, and Howard Suber.

For their guidance during the completion of the book, I thank: the series editors, George Gerbner and Marsha Siefert, at the Annenberg School of Communications, University of Pennsylvania; Rachel Toor, my editor at Oxford University Press; and my agent, Sandra Dijkstra.

My friend and colleague Leita Hagemann worked with me during the initial stages of conceptualization. I am grateful to her for her good ideas, her hard work, and her continued friendship and support.

My sister, Patricia Harriman, assisted me with some of the earliest drafts of the manuscript, offering her enthusiasm and encouragement along with her technical help.

My graduate students were a great help to me all along the way. Special thanks go to: Cheryl Chisholm, Janice Drickey, Sonya Hamilton, Diane Heller, Ha-Il Kim, Lynne Kirby, Dan Koestner, Evelyn O'Neill, Lynn Spigel, and Richonda Starkey.

Throughout the writing of this book, my dear friend Peter Broderick has been invaluable. He devoted endless hours of editing and re-editing, cheered me up during those inevitable moments of discouragement, prodded me from time to time, and above all, believed in me.

I am very grateful to my other friends and family who saw me through this long, and sometimes painful process: Jessica Frazier, Irene Harwood, Natalie Kahn, Jonathan Kuntz, Suzanne Regan, Susan Ridgeway, my parents Robert and Ellen Rose, my brother Rob Rose, and my sister Julie Payne.

And finally, I offer my deep appreciation to my husband, Jeffrey Chester, for his critical insights and for his love and support.

Contents

TARGET: PRIME TIME

CHAPTER ONE

Prime Time as Political Territory

More than 100 million Americans waited in suspense before their television sets. They were about to witness a media event. It had been discussed daily—in the papers, on television, in church, and at school. Psychologists had warned of its dangers, urging parents not to let their children watch alone. Switchboard operators were poised for a blitz of telephone calls. But this wasn't a moon landing or a major sporting event; it was a television movie. On November 20, 1983, American viewers braced themselves for the three-hour trauma of a fictionalized nuclear attack on the U.S.

In the months prior to its broadcast, *The Day After* sparked unprecedented public debate and political controversy. Scheduled to air one month before NATO's deployment of its first intermediate-range missiles in Europe, the film became a rallying point for nuclear freeze advocates. To promote their cause, the activists orchestrated a national campaign around *The Day After,* involving more than 1000 grassroots organizations. Spending over $300,000, they distributed hundreds of bootlegged videotapes of the movie; created provocative TV spots; set up an 800-number hotline; and ran ads in newspapers and magazines across the country. To generate free press coverage, they held candlelight vigils, political rallies, and marches.[1]

Quick to respond, conservatives mounted a campaign to attack the film. Evangelist Jerry Falwell, leader of the Moral Majority, branded the movie "a threat to our national security"

Scene of nuclear devastation from ABC's *The Day After*. *(Courtesy of ABC Circle Films and the Terence O'Flaherty Collection, UCLA)*

and dispatched 80,000 letters, calling for a national boycott of the show's sponsors. Reed Irvine, head of the right-wing group, Accuracy in Media (AIM), lambasted the movie and demanded equal time.[2]

ABC executives skillfully capitalized on the audience interest generated by the controversy, and managed to assuage critics, by following the movie with a special panel discussion. Supporters of nuclear deterrence (Henry Kissinger and William F. Buckley) faced off against arms-control advocates (Carl Sagan and Robert MacNamara). While the network consistently maintained that *The Day After* was "not a political statement," critics labeled the movie "the most politicized entertainment program ever seen on television."[3]

While more visible than most, the fight over this TV movie was just one in a long series of battles fought by political advocacy groups for influence over prime-time television. These conflicts have periodically transformed the normally placid landscape of prime time into a battlefield. In the war for the American mind, entertainment programs have become political territory. Like Pearl Harbor, Iwo Jima, and Pork Chop Hill, programs are remembered by advocacy groups as decisive battles:

Beulah Land—blacks fought with NBC over the image of slaves in the antebellum South.

Maude— pro-choice and right-to-life activists clashed over abortion in the CBS comedy series.

Soap— angry church groups crusaded against ABC to try to keep this racy new comedy from making its network debut.

Other bitter struggles were fought over: *The Untouchables; Chico and the Man; Jesus of Nazareth; Streets of L.A.; Playing for Time; My Body, My Child; Marcus Welby, M.D.; Policewoman; Cagney & Lacey.*

During the twentieth century, all forms of the mass media have been targets of advocacy groups at one time or another. Motion pictures, comic books, plays, newspapers, and textbooks have been subjected to pressure. But prime-time television has far surpassed all of these as a focus of protest. No other medium has experienced as much pressure from as many organized groups for as long a period of time. As early as the 1950s prime-time television was under attack: from black groups, from anti-communists, and from a host of others who criticized the new medium for commercial excesses and violent content. As the power of television grew, the pressures mounted. By the late sixties, they rose to new heights. In the 1970s this clash of forces turned prime-time television into a "contested zone." By the early eighties, more than 250 advocacy groups had been involved in efforts to change network television. When industry leaders gathered in 1981 to address the problem of "the proliferation of pressure groups in prime time," they declared that such groups "must now be recognized as a permanent part of the television environment."[4]

Why have so many groups taken aim at network television? The reasons are rooted in the special role played by television in our society. In its short history, television has become our most powerful and pervasive mass medium. Virtually no home is without a TV, even if there is no telephone. Families watch television an average of seven hours a day. Television has rapidly permeated both our public and private spaces; as TV sets have multiplied inside the home, they have also moved into hospitals, train stations, airplanes, restaurants, and even parks and beaches.

People rely on television for much of their information about the world. It has surpassed newspapers as the primary source of news for most Americans. In national emergencies, TV is where people turn for information and reassurance. This dependency on television endows it with considerable power—to focus national attention on a single event, to make or break a candidate, and to set the agenda for public debate.

But television's greatest power is in its role as the central storyteller for the culture. It is the fiction programming, even more than news and public affairs, that most effectively embodies and reinforces the dominant values in American society. This is what has made television such a critical target for political groups.

Television, like motion pictures, engages audiences through a powerful combination of narrative and visual images. But unlike theatrical movies, television tells stories around the clock.

The networks, cable services, independent stations, and public television pump an enormous volume of fiction programming into American homes every day. No other mass medium has ever supplied such a large quantity of story material on a continual basis. In the words of Professor George Gerbner, "Television tells most of the stories to most of the people, most of the time."[5]

Prime time draws most of its material from contemporary American life, refashioning it to fit television's own commercial and institutional conventions. Though the patterns of this "world of television" do not match the contours of the real world, the more people consume TV's images, the more likely they are to confuse them with reality. Even those who don't watch a great deal of television find it hard to escape the influence of these images. "We hear about them at work," observes Todd Gitlin, "or from our children, or parents, or friends, or encounter them transfigured into the styles of people in the street. . . ."[6]

Because of this, TV's plots and characters take on special symbolic significance. The contents of prime time have been measured and monitored by advocacy groups, as well as critics and social scientists. Studies have focused particularly on what is missing or misrepresented. They have argued that:

- television depicts unions as "violent, degrading, and obstructive";
- prime time's businessmen are "crooks, clowns, and conmen";
- aging Americans "are portrayed as more stubborn and eccentric than others";
- TV families rarely tackle such real-life problems as "finding and paying for child care and stretching the family budget."[7]

While these critiques point out prime time's failings in specific areas, others have leveled broader and harsher complaints. "Television," charged one advocacy group leader, "is an electronic vending machine, offering sweetsmelling bodies, stuffed, satisfied bellies, and great vacant gaps in our cerebral cortex."[8]

The advocacy groups that have mobilized to change television are difficult to categorize, count, or even keep track of.

They've fluctuated in numbers and intensity, and varied in size and composition. They've been called "pressure groups," "special interest groups," "lobbies," and "citizens' groups"—depending on who's doing the labeling. There have been large groups with enormous resources at their disposal, and small groups operating on a shoestring. For some organizations, television is only one of their targets; others have focused on television exclusively. Sometimes several groups have struggled to represent the same constituency. In other cases, rival groups have teamed up in ad hoc coalitions. While some organizations have remained a presence over the years, others have been as ephemeral as the programs they sought to influence.

They've represented a myriad of causes, and have differed sharply in their perceptions of television and what they wanted from it.

To **minorities, women, gays, seniors,** and the **disabled,** television is a cultural mirror which has failed to reflect their image accurately. To be absent from prime time, to be marginally included in it, or to be treated badly by it are seen as serious threats to their rights as citizens. The National Organization for Women, the National Association for the Advancement of Colored People, the National Gay Task Force, and a host of other groups have campaigned for fuller and more positive representation in the world of television.

To **conservative religious groups,** television is a threat to traditional values, too often a dangerous intruder in the home. Organizations including the Moral Majority and the National Federation for Decency have pressured the TV industry to stop the "tide of degeneracy" which they believe threatens to engulf the American family.

Social issue groups believe television is an electronic classroom, in which lessons are taught by the heroes of prime time. Groups such as the Population Institute and the Solar Lobby point to incidents of instant impact—like the time Fonzie of *Happy Days* got his first library card and inspired thousands of youngsters to do likewise. A number of groups have sought ways to incorporate messages about birth control, drug abuse, nuclear war, and a range of other issues into the plots of TV's nightly entertainment.

Anti-violence groups see television "murder and mayhem" as a toxic substance. They fear that continued exposure to this poison will produce a more violent society. The American Medical Association, the PTA, and other organizations have tried to force the networks to reduce TV violence.

However their perspectives and goals differ, all of these groups confront the same institution. While television programming is a familiar and constant presence, its processes are distant and cloistered. Because of its size, power, and complexity, network television seems impenetrable. The high-level decision making that determines what America sees is shrouded in mystery. Prime-time programming is created and distributed through a complicated web of Hollywood studios, advertising companies, and local stations. At the center are three powerful New York-based corporations.

There appears to be no way for the public to influence what goes on in this web. While patrons of movies and theaters can cast their votes at the box office, in television, a handful of anonymous Nielsen families have voting power for the whole nation. "Many Americans," notes critic Howard Rosenberg, "share a sense of alienation from an unreachable, untouchable institution that has great influence on their lives."[9]

But no matter how powerful and remote network television may seem, it has its own unique vulnerabilities. Television has always been a federally regulated medium. While the network corporations themselves are not licensed, their stations are. Governmental oversight—through the Federal Communications Commission and Congress—has often forced network television to pay attention to public pressure. And advertisers, because of their vulnerability to boycott threats, have sometimes been called "the soft underbelly of the networks."

Though network television may appear monolithic, it is made up of a complex collection of organizations, each with its own unique subculture. While highly interdependent, they do not always share the same interests or concerns. Decision making in television can be, in Muriel Cantor's words, a "negotiated struggle." Some attacks by advocacy groups have been like lightning bolts, illuminating internal divisions among the networks, studios, stations, and advertisers.[10]

All the groups that have tried to influence prime time have drawn upon an arsenal of weapons designed to strike at the characteristic "pressure points" in network television. These weapons have been used in very different ways, depending on the strategies of each group, which in turn were determined by their goals, perceptions, and levels of sophistication.

The most ad hoc strategy for pressuring a network is a national **protest.** In such protests, groups have employed every possible weapon at their disposal. They've threatened sponsor boycotts, pressured network affiliates, launched massive letter-writing campaigns, and taken their battles to the press. Another strategy has been to **appeal to governmental authorities,** by petitioning Congress, filing civil suits, and complaining to the Federal Communications Commission. Some have **lobbied** the television industry, offering information and incentives to encourage incorporation of their issues into prime time. Still others have launched **campaigns**—well-orchestrated, highly publicized, planned attacks on network television, designed to force broad changes in its programming.

Clashes between advocacy groups and network television have often sparked heated debates. Some observers have argued that these groups are a threat to First Amendment freedoms and that their efforts to influence programming content restrict free speech. Commented Carrie Rickey: "Liberals charge that this Fire on the Left and Brimstone on the Right threaten every American's inalienable freedom of speech. Let the civic or federal government interfere with the making of a movie, writing of a book, or placement of a sculpture (no matter how odious, how offensive) and you have censorship, that frightful progeny of a prescriptive authoritarian government." Producer David Gerber warned that "pressure groups" trying to "implant their own thinking on television" will turn the medium into a "tower of Babel." CBS's Gene Mater predicted that "special interest groups," each with their various and sundry demands, would ultimately "balkanize" network television.[11]

Others have maintained that these groups play a useful role in a pluralistic society because they ensure that various interests and issues are represented in the mainstream media. Producers Richard Levinson and William Link have suggested that

". . . pressure groups are for the most part a necessary goad. Without their complaints, strident or otherwise, the television community would perhaps fall victim to its own parochial interests." Some commentators have argued that the First Amendment protects the free speech only of those who control the media. They have asked, "What recourse do disenfranchised groups have but to protest?"[12]

The chapters that follow trace the history of advocacy group involvement with network television. After describing the origins of media activism in the fifties, and its rise in the sixties, the book focuses on the most dramatic and significant cases from the last two decades. Political shifts, changes in government policies, and market fluctuations all have affected significantly the power of advocacy groups. The interplay between the groups and the networks has been dynamic; pressure strategies have been met with industry counter-strategies. At various times, the networks have negotiated, cooperated, or capitulated. These conflicts have been reflected on American television screens. Sometimes the impact has been dramatic; often it has been subtle but significant.

Each of the stories in the following pages presents a different facet of a complex mosaic. Seen together they present a full picture of the most powerful medium in the world under pressure.

CHAPTER TWO

Television Under Siege

In January 1969, a press conference was held at the St. Regis Hotel in New York. Leaders of the newly formed National Citizens Committee for Broadcasting (NCCB) announced a plan to combat "airwave pollution." The group called for a "stop to the broadcast-government liaison that in the name of free enterprise has exploited audiences—our nation's people—and enriched a handful of their peers." Proclaiming itself "a voice for the people, who have had no voice in how the public airwaves have been used," NCCB threatened to challenge the licenses of stations around the country "at random."[1]

NCCB was only one of the many activist groups campaigning against broadcasters in the late sixties and early seventies. Citizens were organizing all over the country to reform the television industry. Never in its history had broadcasting experienced such an onslaught. Television was under siege. The reformist groups were serious in their purpose and committed in their efforts. They used every weapon at their disposal to force changes in television: marches, sit-ins, pickets, and challenges to broadcast licenses. Like the lead character in the movie *Network,* they were "mad as hell and not going to take it anymore!"

These pressures did not spring up overnight. They originated in an inherent contradiction at the very foundation of American broadcasting. As early as the 1920s, Congress had designated the airwaves to be public property. In exchange for the privilege of using scarce frequencies, radio stations were expected to offer programming that served their communities.

But the imperatives of business inevitably drew broadcasters away from their public mandate. A small number of national networks quickly dominated the system. As network radio became a national market for products, programs were designed as vehicles for advertising messages. Advertising—not the public interest—became the driving force in radio, influencing the formats as well as the content of programming. The same corporations that controlled radio were the founders of television. Following the profitable radio model, television developed into a merchandising medium.

It was this conflict between public responsibility and private profits that eventually transformed television programming into political territory. Though the origins of the clash were there long before the first TV broadcast, it took a series of developments in the fifties and sixties—both within the television institution and in the society at large—to put television on a collision course with the public.

Television offered a variety of programs during its first few years. With long hours to fill in the programming schedule, there was room for experimentation. In addition to popular situation comedies and variety shows, viewers could see live drama, public affairs, and cultural programs. The system of TV program procurement and scheduling was decentralized, with advertisers packaging most of the programming in prime time. "An independent producer," explains J. Frank Reel, "could produce and sell television programs in a free and open market. There were a large number of potential customers for a television series. A producer could sell it to any one of dozens of national advertisers for network exposure. . . ."[2]

Sponsors had direct control over program content. Advertising messages were often integrated directly into the story line. Sponsors reviewed all scripts and story ideas, deleting any material that might reflect negatively on their products. This direct relationship between sponsor and product made it possible for outside groups to pressure the television industry by threatening sponsor boycotts. Anti-communist groups effectively employed this tactic as part of a sophisticated campaign to blacklist alleged communists in the television industry.[3]

Laurence A. Johnson, who owned a small chain of supermarkets in upstate New York, devised a unique method of pressuring sponsors. He placed signs next to certain products on his grocery shelves, warning customers that purchasing these goods would advance the spread of communism. To ensure the cooperation of the advertising industry, he made regular trips to New York City, traveling up and down Madison Avenue to advise advertisers about whom to hire and whom not to hire. Most of the companies willingly cooperated with the system, justifying their action as good business.[4]

The success of this blacklisting campaign highlighted the vulnerability of a system of advertiser-supported programming. Advertisers could easily be intimidated by political groups if product sales appeared to be at risk. Such a system was bound to make the broadcasting industry a potential target for pressure. The tactic of pressuring television advertisers would be used by other groups. As the relationship between sponsors and programming became less direct in the sixties and seventies, the task of boycotting advertisers would prove more difficult. Newer, more sophisticated techniques would be devised.[5]

Failed Promise

Black groups were less successful in their attempts to protest *Amos 'n' Andy* in 1951. Billed by the TV industry as the first "all black network TV show," the CBS series was condemned by the 350,000-member National Association for the Advancement of Colored People (NAACP) for depicting "Negroes in a stereotyped and derogatory manner." The civil rights group tried every possible tactic to get the series off the air: a letter-writing campaign, a lawsuit, and a nationwide boycott against the show's sponsor. But their efforts failed. Though the sponsor—Blatz Beer—did finally withdraw, the way this particular series was packaged made it less susceptible to economic pressures from blacks. Unlike most other network shows during this period, which were produced by advertisers, *Amos 'n' Andy* was owned by CBS. When one advertiser withdrew support, another could replace it. This made it possible for the network to shift the controversial series into syndication, selling reruns not only to individual stations in the United States, but also to television systems around the world, including England, Austria, Kenya, and Western Nigeria. *Amos 'n' Andy* remained on the air for the next fifteen years.[6]

Black activists were particularly disturbed at their powerlessness in keeping the offensive series off the air, because they had looked on this new medium with high hope. With its need for new talent, television presented seemingly unlimited possibilities for employment and for wide exposure to the American public. Television was hailed as the new frontier for blacks, a medium "free of racial barriers." Ed Sullivan, host of CBS's variety show, *Toast of the Town,* wrote in his weekly newspaper column: "Television not only is just what the doctor ordered for Negro performers; television subtly has supplied ten-league boots to the Negro in his fight to win what the Constitution of this country guarantees as his birthright. It has taken his long fight to the living rooms of America's homes where public opinion is formed and the Negro is winning!"[7]

But, as the *Amos 'n' Andy* case dramatically illustrated, in a profit-driven medium, commercial success and social responsibility could often be at odds with one another. Though blacks did find some opportunities in television, especially in the first few years, they were most likely to appear in variety shows as dancers and singers, or in a handful of comedy series where they were cast in stereotypical roles. While the networks paid lip service to the black cause, their actions were governed by the imperatives of the broadcasting business. NBC hired a public relations firm to promote "a more realistic treatment of the Negro on the air and the hiring of more Negro personnel." But featuring blacks in prominent roles or addressing black issues directly in prime-time programming was considered too risky. Networks were easily influenced by southern affiliates who refused to carry shows that presented blacks too positively. The networks needed to have their programs carried in as many cities as possible to ensure that advertisers reached the large, national market that network television promised to deliver.[8]

Consequently, safe, minimalist policies were devised to assuage blacks by providing them some employment and visibility. NBC's policy of "integration without identification" marginalized and circumscribed the appearance of blacks in prime time. As an NBC executive proudly explained in 1957:

> It is so easy, really, in casting for sympathetically-portrayed roles to hire actors whose racial derivation is apparent. . . . I hope you have noticed here and there everything from taxi-drivers to newspapermen, from doctors to social workers, played by competent Negro actors or actors of other racial minority derivation.[9]

These measures only postponed until the next decade more serious confrontations between black activists and the television industry.

Rising Crescendo of Criticism

During its formative decade, television entered American homes at a dramatic rate. In 1951 only 10 percent of U.S. households had televisions; by 1961 over 90 percent owned sets. Television had become the country's primary mass medium. At the same time television was rapidly intruding into American homes, economic forces were shaping it into a strongly centralized, highly commercial medium. Efforts to create regional networks failed, and by the mid-1950s, three national TV networks dominated the television system. As audiences increased, broadcast time became more valuable. To maximize these profits, the networks began to take more control over their schedules. Programming formats designed to garner the largest audience for the least amount of money began to dominate. Filmed series, which could be repeated indefinitely, replaced costly, ephemeral live production. Explains media historian James Baughman:

> By the end of the 1950's, most of what had distinguished early television had all but disappeared. Programs like *Omnibus* and *See It Now* left the air; congressional hearings were no longer aired extensively. . . . The networks also sharply reduced the number of spectaculars and dramatic anthologies, while the surviving ones originated not live from New York but on film from Hollywood. More of the schedule, all told, went to those program types already on the air—situation comedies and action

> melodramas—most with a standard cast and setting. Little room remained by 1959 for that which could not be quickly constructed on a studio assembly line.[10]

These trends were greeted by a "rising crescendo of criticism" over television's "middlebrow, mediocre programming." Groups like the National Association for Better Radio and TV, the National Audience Board, and the PTA publicly urged television to improve. Stronger complaints were voiced by educators, parents, and writers. *Parents* magazine urged its readers to "get rid of tele-violence." *The New Yorker* accused television of turning performers into hucksters. Author John Steinbeck charged that the medium had "taken the place of the sugar-tit soothing syrups, and the mild narcotics parents in other days used to reduce their children to semi-consciousness and consequently to semi-noisiness." And a *Reader's Digest* article labeled television "a ghastly neurosis" and offered "Audiences Anonymous" as a cure.[11]

The specter of tighter federal regulations loomed over the network industry throughout its first decade, becoming more ominous during times of public criticism. Television was kept in the government spotlight by periodic congressional investigations into TV violence and television's contribution to juvenile delinquency.[12]

The television industry responded by developing mechanisms to deflect criticism. To ward off regulatory intrusions, a "Code of Good Practice" was established by the industry's lobbying and self-regulatory body, the National Association of Broadcasters (NAB). As a result of a skillful public relations effort by the NAB, articles about the Code appeared in *TV Guide* and other popular magazines throughout the fifties. But the Code was ineffectual: the content guidelines were purposely vague; a small board was responsible for monitoring hundreds of stations across the country, and there was no real punishment for violators. Nevertheless, the Code served as a useful shield for the industry. Whenever public or government criticism increased, the Code was presented as evidence that broadcasters were already regulating themselves.[13]

A Deep Scar

But the Code was unable to protect the industry from the national scandal that erupted in 1959 when contestants on phenomenally popular quiz shows such as *Twenty-One, The $64,000 Question,* and *Dotto* revealed that the programs were rigged. The quiz shows were abruptly taken off the air, and Congress launched an immediate investigation of the television industry. The congressional committee report expressed outrage at "how far certain advertisers, producers and others will go to wring the last possible dollar of profit out of the privilege of using the airwaves." The report called for sweeping legislation to strengthen the powers of the FCC over the networks.[14]

The networks acted quickly to pre-empt government action. At the same time, they managed to increase their control over programming. Sponsors—whom public opinion held largely responsible for the quiz scandals—were phased out of their direct packaging of programming. Network policies shifted from single sponsorship to the "magazine concept," where advertisers would buy commercial "spots" much like they did in print media. These changes accelerated trends already under way in an industry where rising costs for broadcast time were making it too expensive for advertisers to buy entire programs.[15]

To prevent any more embarrassing incidents, all three networks strengthened the power of their continuity and acceptance departments. As the official network censors, these departments had been part of the network structure since the radio days, developing and administering policies for program content. Now the networks transformed them into more powerful "broadcast standards" or "program practices" departments and gave these units more control over program content, charging them with the responsibility for overseeing the creation, production, and broadcast of all prime-time entertainment programs.[16]

Network television entered its second decade with mechanisms in place for offsetting criticism from the public. But, despite its effort to promote public good will, the industry continued to be a target. The congressional hearings and press coverage of the quiz scandals had focused a piercing searchlight on the entire television industry. Though the quiz shows had vanished from the airwaves, serious questions remained in the public's mind about the integrity and quality of television. There was now "a deep scar in TV's image." Newly appointed FCC chairman Newton Minow publicly chastised industry leaders in 1961, challenging them to

> sit down in front of your television sets when your station goes on the air and stay there without a book, magazine, newspaper, profit and loss sheet, or rating book to distract you—and keep your eyes glued to that set until the station signs off. I can assure you, you will observe a vast wasteland.[17]

This climate of negative public opinion stimulated a number of groups that had been organized to deal with other issues to shift their attention to prime-time television. The newly enlarged network standards and practices departments* were assigned the responsibility of handling these complaints. For the most part, they were dealt with on a routine basis. In a 1961 newspaper article "TV Censor's Job: Listening to All Those Complaints," NBC's head of broadcast standards characterized such groups as petty annoyances. Referring to himself as "the man in the middle," the executive said his job was to prevent "tasteless, outrageous scenes from turning up on television shows" and "to handle complaints and protests from people and groups who feel they have been mistreated by the medium." The article lightheartedly listed the kinds of complaints that faced the network censors:

> Dental groups howl when a TV character expresses dread of the dentist's chair. . . . Bankers and securities dealers don't like it when one is depicted as a crook or embezzler. . . . Librarians' organizations write bitter little notes about the stereotypes of the librarian. . . . and the wine industry became upset when it was implied that muscatel was the beverage most favored by broken-

*Though these departments went by slightly different names at each of the three networks, the generic term "standards and practices" became a commonly accepted label for them throughout the industry, and will be the term used in this book.

> down alcoholics. . . . Sometimes I think there are too many minorities. . . . and I'm not talking about race, creed, or color, or land of origin. I mean doctors, lawyers, plumbers, storekeepers and milliners. And all of them are very easily offended.[18]

When CBS announced plans to air Arthur Miller's award-winning play *Death of a Salesman,* a prominent group of sales executives stepped forward with both complaints and advice for the network. Claiming that the play had haunted the profession for years with its dismal, hopeless view, the group offered suggestions for "improving" it with just a few script changes. Certain lines could be removed, for example, that are "needless anti-selling, anti-business comments that run through the play and add nothing to either plot, mood, or characterization." The show could be further improved if a brief "prologue" were added called "Life of a Salesman," pointing out that "with modern customer-oriented selling methods, Willy Lomans are ghosts of the past."[19]

Some groups used more heavy-handed tactics. Though number one in the ratings, *The Untouchables* was already under fire from congressional committees and TV critics for its violence when Italian Americans charged that the Roaring Twenties gangster series "defame[d] Italians" by "stereotyping them as criminals." The Federation of the Italian American Democratic Organizations of the State of New York—whose board included United States Congressmen Peter Rodino, Alfred E. Santangelo, and Joseph Addabbo—told the press that its action against ABC was a first step toward "cleaning up the TV industry." The Federation launched a boycott against one of the show's regular sponsors, Liggett & Myers. The group distributed 250,000 posters to stores nationwide, and persuaded the owners of 2,000 vending machines to stop carrying L&M cigarettes.[20]

Liggett & Myers withdrew from the series. Frightened ABC executives, and the show's producer, Desi Arnaz, met with Italian American leaders and hammered out a compromise: *The Untouchables* would no longer feature "fictional characters as hoodlums with Italian names"; more prominence would be given to the one positive Italian American lead character in the series; and, to balance the negative Italian American gangsters, new

Italian American characters with positive attributes would be added to the series.[21]

New York Times TV critic Jack Gould found these changes disturbing. Such an agreement "hold[s] latent dangers for the well-being of television as a whole," he warned. "An outside group not professionally engaged in theatre production has succeeded in imposing its will with respect to the naming of fictional characters, altering the importance of a leading characterization and in other particulars changing the story line."[22]

For the networks, however, the concessions served their purpose. The protesters were placated; the show remained on the network for several more years, and survived long after that in syndication; and the changes in character and plot were hardly noticeable to the show's general viewership. Such effective strategies would be used again.

Media Reform

While TV executives were developing techniques to deal with trade associations and protesting pressure groups, political forces were organizing a more broad-based attack on the television industry. Growing out of the black struggle for civil rights, it quickly expanded into a full-scale "media reform movement," involving a wide spectrum of organized groups.

During the same years in which television was becoming America's dominant mass medium, the civil rights movement was gaining momentum. White, liberal organizations joined black groups in their fight to reverse racial discrimination and segregation in American society. These groups were also learning how critically important television could be as a political tool. Network news coverage of the desegregation efforts gave the civil rights cause unprecedented national exposure. With more impact than print media, television brought into American living rooms the drama and violence of the black struggle for equality in the South.[23]

This amount of public exposure in television news contrasted sharply with the absence of blacks from entertainment programming. Industry promises during television's early years had not been fulfilled. In a mass medium that commanded the daily attention of nearly every American, blacks needed full representation if they were going to make gains in society.

Black activists launched a series of campaigns in the early sixties to force changes in TV industry practices. They employed the same tactics which were gaining them access to other American institutions: widespread publicity, appeals to the government, and direct confrontation when needed. To demonstrate their seriousness, activist leaders made repeated threats of economic boycott, picket lines, and legal action. Congressional hearings were held to expose industry leaders for their failure to keep their promises to blacks. The NAACP and the Congress on Racial Equality (CORE) began pushing the film and TV unions to hire more blacks. A major drive was launched to get the advertising industry to incorporate more blacks in commercials and in print. Following passage of the Civil Rights Act in 1964, the New York Ethical Culture Society began a nationwide monitoring campaign to "prod network executives to remedial measures," charging that "equal protection under the law was being denied Negro children through the distorted world television offers them."[24]

By the mid-sixties, visible changes could be seen in network entertainment programming. ABC announced it had "stepped up integration" on daytime soaps. NBC introduced *I Spy*, the first series to have a black costar, comedian Bill Cosby. "This year, as never before," observed one TV critic in 1965,

> television shows are reflecting the Negro revolution. Negro actors are finding work and they are getting roles that could be filled by whites. . . . There is a Negro teacher in *Mr. Novak*, a Negro secretary in *East Side/West Side*, Negro students in the classroom of the *Patty Duke Show* and on the campus of Channing. On *The Outer Limits*, a Negro recently played a would-be Pierre Salinger, press secretary to a presidential candidate. Another *Outer Limits* show will have a Negro astronaut.[25]

By the end of the decade, notes J. Fred MacDonald, "there were more than two dozen network programs on the air featuring black actors as leading characters, or in prominent, regular supporting roles."[26]

While black activists pushed for integration in programming, a separate but related political effort was launched, aimed at more widespread changes in the way the broadcasting industry operated.

In 1964, a coalition of civil rights groups filed a petition with the FCC asking the Commission to deny the license renewal of TV station WLBT in Jackson, Mississippi. These groups claimed that WLBT's owners had blatantly discriminated against blacks in hiring and programming. The move was unprecedented. "Petitions to deny" had been used in the past only by those companies or individuals proposing to use the same broadcast frequencies to start their own stations. They were employed infrequently, and the FCC routinely renewed broadcast licenses. The challenge against this local station was part of a national campaign spearheaded by the New York-based Office of Communication of the United Church of Christ. The goal was to use regulatory mechanisms to force television to be responsive to the public.[27]

To provide support for their case, the petitioners had carefully gathered evidence of the station's discriminatory behavior. Volunteers monitored WLBT's programming. Detailed reports were prepared which documented abuses, such as the interruption of network news coverage of the civil rights movement with graphics saying, "Sorry, Cable Trouble." The station repeatedly broadcast racist editorials without offering the opportunity to reply, and consistently used the words "nigger" and "nigra" in its broadcasts.[28]

When the FCC granted a one-year license to the station without allowing the advocacy groups a hearing, the UCC went to court. In March 1966, the U.S. Court of Appeals for the District of Columbia ruled that the FCC had to grant these groups a hearing. This historic decision meant that the petition to deny could now be used as a weapon against station licensees.[29]

The WLBT case marked a critical turning point. Organized groups had established a legitimate power base for influencing the broadcast industry. As *Broadcasting* noted: "The case did more than establish the right of the public to participate in a station's license-renewal hearing. It did even more than encourage minority groups around the country to assert themselves in broadcast matters at a time when unrest was growing and blacks were becoming more activist. It provided practical lessons in how pressure could be brought, in how the broadcast establishment could be challenged."[30]

Out of this landmark case emerged a movement committed to reforming television. Media watchdog groups began to spring up around the country. Many of them represented those disenfranchised segments of American society that had begun to mobilize in the wake of the civil rights movement: Hispanics, women, homosexuals. These groups demanded that local stations hire more women and minorities, provide programs to deal with social and political issues, and allow groups access to programming decisions.[31]

Several East Coast organizations provided leadership, training, and support for these grassroots groups. The UCC's Office of Communication, with substantial funding from the Ford Foundation, began training hundreds of groups. Activists with little understanding of the television industry were given crash courses on the intricacies of communications law and effective strategies for pressure. Volunteers were taught how to conduct a monitoring campaign, how to file a petition to deny a station's license, and how to negotiate effectively with television executives. Former FCC commissioner Nicholas Johnson offered advice in his guidebook for citizens' groups, *How to Talk Back to Your Television Set*. Groups like the National Citizens Committee for Broadcasting, Media Access Project, and Citizens Communication Center, staffed with recent law-school graduates, offered a variety of free services.[32]

To counter the lobbying efforts of the National Association of Broadcasters, public interest groups began pushing their own program in Washington. They petitioned for rule changes, introduced legislation in Congress, and closely monitored the FCC's behavior. After persistent pressure from the United Church of Christ and others, the Commission issued a new policy in 1968 for "denying radio and television licenses to stations engaging in racial discrimination." Alarmed at the over-commercialization and poor quality of children's TV programming, five Boston women formed Action for Children's Television (ACT) in 1970. They fought for and ultimately won special FCC guidelines for children's television. In 1972, activist groups succeeded in getting Benjamin Hooks appointed the first black FCC commissioner.[33]

In a very short period, the number of media reform groups mushroomed. Station license challenges increased dramatically. In (fiscal year) 1969, two were filed; in 1970, 15 petitions were filed against 16 stations; in 1971, 38 petitions were filed against 84 stations; and in 1972, 68 petitions were filed against 108 stations. Though very few stations actually lost their licenses, the petition to deny became a powerful weapon of intimidation. The rare instances in which the FCC granted short-term licenses as the result of a pressure campaign were chilling reminders to the rest of the industry that no station was really safe, now that political advocacy groups had new legal power. Station managers soon decided it was much easier to meet and negotiate with group leaders before a petition was filed. They frequently agreed to a few concessions in order to avert further trouble. With the help of the FCC, local stations worked out a formal set of procedures that regularized their relationships with activist groups in their communities. To satisfy demands for programming, groups were given access to cheaply produced public affairs programs scheduled in the late-night and early-morning slots when few people were watching. Such a system placated the groups, without cutting into the stations' profits.[34]

Some groups used the license challenge to influence not just local but also network television. Network-owned stations were just as vulnerable to public pressure as any other local broadcast station. Hence, a new "pressure point" had been identified.

The National Organization for Women used this tactic in 1970 to put the networks on notice. NOW planned to take action to change "the derogatory, demeaning and stereotyped images of women presented by broadcast programming and advertising." A special NOW Media Task Force was set up to coordinate the pressure campaign against the TV industry. To force improvements in prime time, NOW challenged the licenses of two major network-owned stations in 1972: WABC in New York, and NBC-owned WRC in Washington, D.C.[35]

Virtually every major ethnic group mobilized against prime-time TV. While the NAACP and CORE continued their pressure, Black Efforts for Soul in Television (BEST) and the National Black Media Coalition (NBMC) launched their own campaigns. The Italian-American League to Combat Defamation was joined by the American Italian Anti-Defamation League. German-Americans, angry at their prime-time image as "spies, mad scientists and babbling fools," formed the German-American Anti-Defamation League. "Other groups have their 'anti' organizations," the leaders charged, "the Italian-Americans even have two, maybe we ought to have one." Americans of Italian descent joined forces with the Polish-American Guardian Society to pressure the FCC to put an end to "insulting portrayals of nationality, racial or religious groups on TV." When ABC scheduled a new series about Brigadier General George Custer, the Tribal Indian Land Rights Association threatened to petition the federal courts for an injunction to keep the program off the air.[36]

By the early seventies, advocacy groups were pressuring network entertainment television in unprecedented numbers. These groups viewed the media as a critical arena for their political struggles. Their most potent weapon against the networks was the right to challenge the licenses of network-owned stations, which had become the Achilles' heels of NBC, CBS, and ABC.

In their frustration and rage, a handful of these activist groups used guerrilla tactics to force network cooperation. Some of them found ingenious ways of getting into network headquarters and taking over executive offices. Others made bomb threats.

While local stations could marginalize advocacy groups in fringe hours, networks had a tougher problem on their hands. Groups were taking direct aim at the portion of the schedule which attracted the largest, most heterogeneous audience and made the biggest profits. They were demanding integration; they were monitoring the programming for negative portrayals; and they were making it clear to the networks that they were not about to go away. Since network executives were unwilling to give up costly air time in order to appease pressure groups, new strategies would have to be worked out.

In the meantime, other forces were further complicating the troubled relationship between entertainment television and the public.

CHAPTER THREE

And Then Came *Maude* . . .

Maude was in trouble. The colorful, outspoken lead character of the popular CBS television series was pregnant and she didn't want to be. At forty-seven, she was too old to start raising another child. The distressed Maude was prepared to accept her fate—until her daughter, Carol, sophisticated, liberated, and divorced, proposed a solution to Maude's dilemma: abortion. "We're free," she assured her mother. "We finally have the right to decide what we can do with our own bodies . . . and it's as simple as going to the dentist. . . . [Y]ou don't have to have the baby. . . . [T]here's no reason to feel guilty and there's no reason to be afraid."[1]

It took Maude two episodes in November of 1972 to make her final decision. But, after much discussion, she did opt for the abortion. In the process, she also talked her husband, Walter, into having a vasectomy. Though Walter chickened out of the operation, the audience was left with the expectation that he would, sooner or later, go through with it. The debate over these critical decisions took place in the homes of more than 50 million Americans. Not all of them were pleased.

Immediately after the first broadcast, the CBS switchboard lit up with 373 angry telephone calls. Within days, the network and the show's Executive Producer, Norman Lear, were deluged with mail from outraged viewers. Among the missives was a packet with a dozen glossy photographs of aborted fetuses. For the next year, *Maude* was embroiled in a major public battle involving political forces on both sides of the abortion issue.[2]

The case was a pivotal one for network television. It tested, as never before, the boundaries of acceptability for program content; it pushed into the public arena a debate about the proper role of television in dealing with controversial political issues; and it had a lasting impact on the way such issues would be handled in future prime-time shows.

Maude was a spin-off of the popular television series *All in the Family,* which had made its debut on CBS the year before. *All in the Family* represented a major breakthrough for entertainment television. The first situation comedy to be built around social and political controversy, the series was also part of CBS's new strategy of designing programs to attract the younger, urban audiences who were becoming more and more desirable to advertisers. The show was, in the words of media scholars Horace Newcomb and Robert Alley, "something of a time bomb." Within three months of its premiere in January 1971, *All in the Family* had "addressed the issues of homosexuality, miscarriage, race, female equality, and cohabitation." Though not an instant success, by the end of its first season, the series had climbed to number one in the ratings.[3]

As the show became more popular, it drew fire from various ethnic groups. Most of the outcries concerned the bigoted lead character, Archie Bunker, whose dialogue was a veritable lesson in racial epithets. Nearly every time he opened his mouth, Archie would spew out words like "spades, spics, spooks, schwartzes, coons, coloreds, and chinks." Such behavior prompted groups like the Anti-Defamation League of B'nai B'rith, the Urban League, and the Polish-American Cultural Society to publicly criticize the show. Lear's response to these critics was that the series was designed to combat racism by making a fool out of Archie and giving his liberal son-in-law, Michael, the last word in their arguments. Whether or not the series succeeded in challenging bigotry became an issue which social critics and scholars debated for some time. But the attention generated by the show helped to enhance its popularity.[4]

All in the Family was in fact so popular that its success dramatically altered prime-time television. The trail-blazing series led the way for a host of new controversial shows appearing in prime time during the fall of 1972. Observed *Time* magazine:

> TV has embarked on a new era of candor, with all the lines emphatically drawn in. During the season that began last week, programmers will actually be competing with each other to trace the largest number of touchy—and heretofore forbidden—ethnic, sexual and psychological themes. Religious quirks, wife swapping, child abuse, lesbianism, venereal disease—all the old taboos will be toppling. . . . An upcoming ABC Movie of the Week will feature Hal Holbrook explaining his homosexuality to his son. . . . NBC's *The Bold Ones* will be getting bolder, mainly by knifing into such delicate surgical issues as embryo transplants and lobotomy. . . . No new adventure hero, it seems, will be admitted to the schedule without an ethnic identity badge. ABC's *Kung Fu* is sort of *Fugitive* foo yung—a Chinese priest permanently on the lam in the American West of the 1870's, nonviolent but ready to zap troublemakers with the self-defense art of kung fu. The title character of NBC's *Banacek* . . . is not only a rugged insurance sleuth but also a walking lightning rod for Polish jokes. . . . Indeed, the 20 new series making their bow this fall add up to a veritable pride of prejudices.[5]

Maude was at the center of the season's bold new shows. The strong-willed, liberated, cousin of *All in the Family's* Edith Bunker, Maude Findlay was played by large, husky-voiced Beatrice Arthur, whom producer Norman Lear described as "the flip side of Archie . . . a Roosevelt liberal who has her feet firmly planted in the forties." The new series promised to take prime-time television even further into controversial territory. Said CBS president Robert Wood: "Maude breaks every rule of television from the start. . . . She's on her fourth husband, and she is living with a divorced daughter who has a son. It's not so long ago that you couldn't show a woman divorced from one husband, let alone three."[6]

While some advocacy groups were complaining about these new topical shows, others saw them as a unique opportunity. One such organization was the Washington-based Population Institute. Set up in 1969, the Institute's mission was to influence public opinion about the perils of unchecked population growth. The group planned to achieve its goal "by increasing the population information content of the news, entertainment and advertising media." The plan to use entertainment television for such educational purposes was the brainchild of two Methodist ministers, David Poindexter and Rodney Shaw. Prime-time programming seemed to be the ideal vehicle for public education. Not only could it reach large numbers of the American public in an instant but it could also package political and social issues in an entertaining context.[7]

In contrast to the groups struggling to gain access to the TV industry, Population Institute leaders were able to go directly to top television executives with their appeals. The issue of population control was already becoming a matter of much public discussion, as legislators, philanthropists, educators, and other public figures lent their names to the cause. In 1971, the Population Institute held a special luncheon at the Waldorf Astoria hotel in New York to persuade TV executives to focus on population issues in their programs. In attendance were such notables as John D. Rockefeller III (who was chairman of the President's Commission on Population Growth and the American Future), Senator Robert Packwood (R-Ore.), and George Bush (then chairman of the Republican National Committee). The top executives from all three networks were also present, including CBS vice chairman Frank Stanton. So was Norman Lear.[8]

As David Poindexter remembers, the three networks agreed at the end of the meeting to show "quality programs dealing with aspects of the population issue." The Population Institute also enlisted the support of the National Academy of Television Arts and Sciences as "a sponsor of TV population education." In the winter of 1971–72, the Institute conducted two-day conferences in Los Angeles and New York at which "the leading creative people in the TV industry discussed with population experts means whereby population education might be advanced and attitudes changed." As an added incentive, the organization offered prestigious awards and cash prizes to writers and producers of TV shows dealing with population matters. Pamphlets were distributed announcing that $10,000 would be offered for "the best prime-time entertainment program of 60 minutes or longer," $5,000 for the "best half-hour prime-time entertainment program," and $5,000 for the "best daytime serial episode or series of episodes, during the 1972–73 season."[9]

The Population Institute's efforts to educate Hollywood played an important role in the genesis of the *Maude* abortion episodes. The organization succeeded in stimulating Norman Lear's personal interest in doing a show about the population issue. In addition, as the producer told the *New York Times*, "half a dozen writers came in with late-life pregnancy ideas" during 1972. The original concept for the *Maude* episode was to have her best friend become pregnant. But, said Lear, "I realized the only way to engage the audience's interest was to let Maude get pregnant." Once that decision was made, Lear saw only one way for Maude to solve her problem. As he recalls: "When we were working on the story, it became clear to me that there was no way I could let it be a false pregnancy, or have an accident that caused the baby to be aborted. . . . So then I knew we had on our hands the difficult job of dealing with abortion fairly and getting it on the network."[10]

While prime-time programs were produced by outside production companies, the networks wielded considerable control over program content, a power they had steadily gained since the quiz scandals. Standards and practices departments generally required that all scripts be submitted for review far in advance of their filming dates. Norman Lear's shows, however, enjoyed a degree of independence from the standards and practices department that was unusual in network television. The producer had fought with CBS executives from the very beginning for control over the content of his own programs, threatening to quit if he could not have his way. Lear's series did not entirely escape the scrutiny of the censors. But because the shows had become so valuable to CBS, because he had strong support from the president of the network, and because Lear was inclined to take his case to the press, the producer had managed to free his shows from some of the constraints that were normally placed upon entertainment programming.[11]

This autonomy was also a product of the unique production method employed by Norman Lear. Unlike the typical comedy fare on network TV at the time, *Maude* and *All in the Family* were not filmed, but were videotaped before a live audience. This meant that each episode was on a tight production schedule, with a half-hour episode taped each week. Very often the scripts would not be submitted to the network until the episode had already gone into production. Then changes were worked out during the rehearsal and production process.[12]

But this particular episode was a special case. Lear knew that because the topic of abortion was so politically sensitive, he would at least need to clear the idea with the network before going ahead with it. Months before the broadcast, Lear had a number of discussions with the vice president of standards and practices. As the producer remembers: "CBS was anxious about it. . . . I remember a lot of back and forthing about whether we'd do it and how we'd do it. More about how we'd do it because the network was concerned appropriately that we'd be even-handed since it was a controversial issue." Because the story took a strong pro-choice stand, the network insisted that Lear make special changes for the purpose of balance. A device borrowed from journalistic tradition, "balance" was one of the content policies that standards and practices departments had begun to develop for handling controversial material in entertainment programs. Though at this time the policy was still in its formative stages, in the following years it would evolve into a rather elaborate formula, employed routinely to pre-empt criticism.[13]

To provide balance in the *Maude* episode, a peripheral character was inserted into the story. A heavy-set, cheery woman about Maude's age stopped in for a moment to visit the Findlay family. Already the mother of four rambunctious children, she also had accidently become pregnant again. But, in contrast to Maude's distressed state of mind, this woman was ecstatic about her condition. As Lear later described the character: "She did not need another child, probably would not have wished another child. . . .[S]he didn't need to have another kid yakking at her. And yet in the sunniest way, she said she was excited about having the baby and would never think of not having that child. There was a big effort to get that exactly right." This character never really entered the discussions about the abortion issue; she was merely added as a kind of sidebar to the central story-line, and a very minor one at that. Needless to say, her presence did little to prevent the show from attack.[14]

Even with a character added for balance, the two episodes of the comedy series were pretty strong stuff for network television. Network executives were alarmed when they received the script. Though Lear had discussed the story line by phone with network censors, he had followed his typical routine of not letting the network see the script until production time. Consequently, there was no time to make changes. In a heated exchange of memos, CBS executives reprimanded the producer for his tardiness in submitting the scripts, and Lear retorted that the production schedule made it impossible to turn them in any earlier.[15]

CBS then tried to postpone the controversial episodes to allow time for revisions. But Lear countered with a tactic he had used before to get his way with the network. If CBS refused to go forward with the abortion segments of *Maude* as planned, the producer threatened, then "the network would have to get itself some other show to fill *Maude's* 8 p.m. Tuesday spot on the CBS schedule."[16]

The network was not alone in its concern over the show's reception. Population control groups, who had closely followed the development of the two episodes, braced for trouble. Although the Population Institute had provided the incentive for the two episodes, it had not served as a technical consultant on the script. That role was played by the Los Angeles chapter of Planned Parenthood, which was easily accessible to Hollywood writers and producers. As a result, Planned Parenthood leaders were in a good position to anticipate opposition from anti-abortion groups and to plan a counter-strategy. In an "action memo," Planned Parenthood's Information and Education Office warned member groups that a "barrage of objections from anti-abortion groups [is] expected." To prevent such protests from keeping the second part of the episode off the air, members were advised to write "as individuals" to the local network affiliate stations, to the head of standards and practices for CBS, and to the FCC.[17]

Network affiliates were not required to carry all of the programming fed to them by the networks. Though these local stations routinely aired the network's shows, they occasionally pre-empted them with other programs, especially in cases where the network programming might cause trouble for the stations in their local communities. Stations had always been sensitive to local pressures, but had become even more so with the recent rise in media activism. Anticipating that some stations might balk at airing the two *Maude* episodes, pro-choice groups around the country took steps to guarantee their broadcast. When a Detroit station decided not to carry the show, a coalition of groups pressured the station manager to change his mind. As the Population Institute's David Poindexter explained a few days later in a letter to Lear: "A small task force of us from our Center, from Planned Parenthood-World Population and from some of the Protestant churches got on the long distance phone and stirred up a considerable bit of protest in Detroit, with many leaders in Michigan getting on the phone and speaking to Mr. Carino at WJBK-TV." As a result, the station's management reversed its position and agreed to air the controversial segment. In another case, the activists were not as successful. Two CBS affiliates in Illinois—one in Peoria and one in Champaign—refused to broadcast the *Maude* episodes. The station manager in Champaign justified his action on the basis that the show might be in violation of an Illinois state law which forbade advertising, encouragement or advocacy of abortion. Though the local chapter of the National Organization for Women filed a class-action suit against the affiliate and tried to get an emergency court injunction requiring the station to air the show, the circuit court refused to grant it to them.[18]

Few people involved with the show anticipated the extent of the outcry over Maude's abortion. Hundreds of letters continued to pour in to the network and the production company in the weeks following the broadcasts, most of them highly critical. The major complaint from viewers was that it was inappropriate to treat such a serious subject as abortion in a comedy series, especially in a show broadcast at 8:00 p.m., when many children would be watching. Even some who supported the pro-choice position questioned the judgment of the show's creators. "Why did the *Maude* people do it?" wrote one viewer. "What were they trying to prove? That they are honest and brave and can joke about any subject? Abortion isn't funny. Vasectomy isn't funny. They are subjects of deep moral concern to many people."[19]

Predictably, the most organized and vehement protests came from organizations affiliated with the Catholic Church. These groups had been battling the pro-choice movement for some time, and they had a strong stake in the way the highly charged issue of abortion was presented in the media. To them, this latest incident was an unfair use of media power. Caught off-guard by the two episodes, their immediate reaction was shock and outrage. On November 21, the day of the second broadcast, Monsignor Eugene V. Clark of the Archdiocese of the New York Office of Communications fired off an angry letter to Richard W. Jencks, president of the CBS Broadcast Group, accusing the network of "open propaganda for abortion and vasectomy." A few days later, a delegation from the Catholic Archdiocese of Rockville Centre marched to CBS headquarters in New York City and proceeded to blockade the limousine belonging to CBS vice chairman Dr. Frank Stanton.[20]

The strategy used to protest *Maude* drew upon the methods employed by the liberal media reform movement. Like other groups taking on the television industry, the Catholic organizations looked for pressure points within the institution and turned to the government for help. Since the offending broadcasts had already aired, they sought remedial measures to correct the unjust treatment they believed they had received. While the petition to deny was not appropriate in this case, organizers chose the FCC's Fairness Doctrine as a legal weapon. A policy that applied primarily to editorials and public affairs programs, the Fairness Doctrine required broadcasters who presented one side of a controversial issue of public importance to see to it that contrasting points of view were also aired. Advocacy groups had used the Doctrine on a number of occasions to get response time on television stations. More recently, some groups had even asked the Commission to apply the doctrine to entertainment shows. The National Organization for Women unsuccessfully tried to use the Fairness Doctrine in its campaign against the NBC network, arguing that the network-owned stations had violated the doctrine by presenting unbalanced and stereotypical depictions of women in their programming. While NOW's leaders had difficulty proving that the representation of women was a controversial issue, the *Maude* case seemed to provide a better opportunity for a challenge to the network.[21]

As a first step, the National Council of Catholic Bishops filed a formal protest with the CBS network, asking for equal time to present the other side of the abortion issue. Knowing that broadcasters often met their Fairness Doctrine obligations by scheduling opposing point-of-view programming at times when few people would see them, the Bishops warned the network that they would not "settle for a documentary at three o'clock in the morning." They demanded instead that the network devote two additional episodes of *Maude* to programs "supporting the right to life of unborn babies," specifically, "a sequel in which Maude is pregnant and has a baby." If the producers of *Maude* were unwilling to make the sequels, then pro-life advocates demanded they be allowed to produce their own two shows and that the programs be broadcast in *Maude's* prime-time slot. When CBS refused to comply, the protesting groups took their complaint to the FCC.[22]

Though the protesters' demands were extreme, the case itself raised an important issue about the role of entertainment television in society. In its reply to the Fairness complaint, CBS made an argument that would be echoed in ensuing years whenever TV comedies and dramas were involved in political controversy. The *Maude* episodes, asserted the network, were "solely intended for entertainment and not for the discussion of viewpoints on controversial issues of public importance." But the case was not that clear-cut. In their efforts to influence the content of prime-time shows, the pro-choice groups, like other advocacy groups, saw the programs as a good deal more than entertainment. And audiences were no doubt acquiring knowledge and maybe even opinions about the plethora of social and political issues that had begun to permeate the network schedule. The line between public affairs and entertainment programming was becoming blurred. As Richard A. Blake wrote in the Jesuit publication *America:*

> The problem is not controversial content, but the mode of treatment; there are distinctions among the different genres. *Maude* is a comedy; it does not present a discussion of abortion by experts, offer the editorial position of a station, which by law must be identified as editorial and which may be answered by an opposing view under the equal time provision of the Fairness Doctrine, nor finally, does it present a serious dramatic conflict in which a woman faces a tragic decision. Unless I am very much mistaken any of these formats would meet only marginal objection. In "Maude's Dilemma," abortion is not a matter of life and death, it is a joke with a deadly message: the divorced daughter tells Maude to outgrow her childhood hang-ups, since repugnance for abortion is a silly old-fashioned idea.[23]

It would take several months for the Commission to make its final decision on the *Maude* case. In the meantime, anti-abortion groups organized a grassroots campaign to sustain the pressure effort. An existing structure was already in place that made it possible to coordinate a broadly based, nationwide crusade. The Catholic press was used to rally its more than 48 million church members to action. Readers were urged to "write to the Federal Communications Commission and ask for a public hearing into the Maude situation. . . . For every letter to the FCC, why not write another letter to your Congressman and Senators?" Church leaders also called for a letter-writing campaign and a retaliatory boycott against the eight corporations whose ads had appeared in the controversial episodes. The *Maude* show was lambasted in Sunday sermons, and leaflets with titles such as "Battle Plan for Action" were handed out to parishioners, directing them to call or write the local CBS station manager and tell him "that you will not allow your children to watch channel X as long as they continue to carry *Maude*." If church members needed more help putting pressure on local stations, they were advised to buy a paperback copy of the media reform movement's bible—FCC Commissioner Nicholas Johnson's *How to Talk Back to Your Television Set*.[24]

As the protesters against Maude were orchestrating their letter-writing campaign, the supporters of the show countered with their own. They were assisted by the public relations firm of Rogers and Cowan, which Norman Lear had hired to rally support for the besieged series. Both the Population Institute and the national office of Planned Parenthood participated by mobilizing their supporters to write to CBS's standards and practices department and to the FCC. To counteract the pressure being placed on the show's advertisers, pro-choice group members were asked to write letters to those sponsors, commending them for their support of the show and urging them not to withdraw their advertising from the series.[25]

Most advertisers during this period used form letters to respond to complaints about network programming. These letters explained to viewers that advertising companies merely bought time in programs and no longer had any influence on the program content. Because of the size and intensity of the *Maude* protest, however, the targeted advertisers were more serious in their response to the letters they received. The Breck company, for example, wrote a viewer: "Because we make every effort at Breck to place our advertising on desirable shows, we particularly appreciate hearing from you. As you may be aware, we have received some unfavorable comments concerning the show. We plan to forward your letter to the people at CBS responsible for program content."

Given the widespread opposition generated by the *Maude* episodes, many pro-choice advocates feared that network television would never again deal with population control issues. A more positive strategy was needed to encourage the entertainment industry to continue to treat issues such as birth control, abortion, and population control in its programming. Norman Fleishman, of the Los Angeles office of Planned Parenthood, decided to hold a big event in Norman Lear's honor. A number of dignitaries met at the home of UCLA Chancellor Charles Young (whose wife was on the Planned Parenthood board) in a dramatic gesture of support for the producer. Remembers Fleishman: "Lear told his story to these people, told them about the flack and so on. It was a marvelous meal, a kind of a gala thing." The event was so successful, according to Fleishman, that he decided to hold these evening gatherings on a regular basis, as part of an ongoing program to encourage the Hollywood creative community to deal with the population issue.[26]

The intense flurry of letter-writing, public debate, and boycott threats began to die down within a few months of the controversial broadcasts. The FCC case was still pending. But beyond that, the intensity of the campaigns, both against and in support of the show, had declined dramatically. As with other protests by advocacy groups over network programs, the ephemeral quality of television programming made it difficult for momentum to be sustained.

In the meantime, other groups were putting pressure on the networks in response to the provocative, "relevant" programming in that year's prime-time schedule. One of them was "Stop Immorality in TV," a right-wing organization based in Warrenton, Virginia. The group, whose letterhead included Phyllis Schlafly, anti-communist crusader Dr. Fred Schwartz, and comedian Red Skelton, launched a mass mailing campaign, warning followers that "Judeo-Christian principles are being destroyed through television programs that defy the standards we hold sacred." Cited as prime-time offenders of such values was a laundry list of the new controversial shows of the 1972–73 season. *Maude* was at the top of the list. Also included were *Love, American Style; Laugh In; The Dean Martin Show; All in the Family;* and *M*A*S*H.* "Recent programs on ABC's *Owen Marshall, Counselor at Law,*" explained the letter, "have dealt with lesbian seduction and wife-swapping. Recent episodes on *The Bold Ones* and *Marcus Welby, M.D.* have shown homosexuality in a favorable light. . . . These terrible things are happening because good people are not doing anything," the letter said. "In the past, too many people have felt that their only responsibility was to lead a good moral life for themselves. But things have gotten so bad that the time has come when all good people must take positive action to stop this immorality." The letter provided a list of "some of the products that are advertised on these programs, the companies that make them and the person to write to for action," along with sample letters to write to those companies and to the chairman of the FCC.[27]

Prime-time programming was generating protests from other organizations as well. *Bridget Loves Bernie* had been introduced on CBS during the same fall 1972 season as the *Maude* series and, like so many shows that year, was breaking TV taboos. As described by *Time,* the series featured "a well-heeled Catholic girl who falls for a poor Jewish cab driver. In last week's first episode," the magazine wrote, "they got married and promptly gave birth to dozens of Jewish-Catholic in-law gags." Though CBS had employed Jewish and Catholic religious advisers on the series, network executives apparently had not anticipated the pressure they were to get from Jewish clergy. While Catholic leaders—already embroiled in their fight against *Maude*—did not take action against *Bridget Loves Bernie*, "leaders of virtually the entire spectrum of American religious Judaism," noted the *New York Times*, "have asked the Columbia Broadcasting System to withdraw the program on the ground that it makes intermarriage look 'mod' and thus mocks a basic teaching of Judaism." The entire premise of the show was unacceptable, claimed the protesters, because it "treats intermarriage, one of the gravest problems facing Jews today, not only as an existent phenomenon but one that should be totally accepted . . . and it does so in a cavalier, cute, condoning fashion. . . ." Although the series was one of the most popular of the season—number five in the ratings—CBS announced at the end of March, in the middle of the flurry of protest, that *Bridget Loves Bernie* would not be renewed for the next season. In a typical statement of explanation for the cancellation, CBS president Robert Wood told the press that the decision to drop the show was "absolutely removed, independent, and disassociated from criticism of the show from some Jewish groups."[28]

CBS found itself in the middle of yet another controversy that same spring. *Sticks and Bones* was a hard-hitting drama about the bitter experiences of a returning war veteran. The TV movie was scheduled for broadcast in March of 1973. Shortly after the air date was announced, the United States pulled out of Vietnam. Due to pressures from affiliates, who believed the show would be too upsetting at a time when veterans were returning home, CBS postponed the broadcast. The producer, Joseph Papp, outraged at the decision, canceled his four-year agreement with CBS, charging the network with censorship. When the show finally did air in the summer of 1973, it was without any commercials at all. Only 94 of the 200 network affiliates carried the broadcast. Many praised the network for its willingness to broadcast *Sticks and Bones* under such circumstances, but the incident was troublesome and costly for CBS.[29]

The 1972–73 season had become a critical test year for determining just how far entertainment television could venture into controversial territory. As the leader in prime-time controversy, CBS found itself in the spotlight. In the spring of 1973, the network faced the decision of whether or not to rebroadcast the two controversial *Maude* episodes during the regular summer reruns of the series. The safest choice was to forget the reruns altogether and simply let the matter rest. The programs had caused the network enough trouble already. With all the other pressures it was receiving, further controversy was hardly needed. According to David Poindexter, CBS's first decision was indeed, not to rerun the episodes. But the Population Institute stepped in again, this time to prod top network officials to change their minds. As Poindexter remembers, it didn't take a great deal of effort.

After hearing from Norman Lear in April that CBS had opted not to rebroadcast the abortion/vasectomy episodes, Poindexter decided to pay a visit to CBS vice chairman Frank Stanton. Poindexter reminded the network chief of the commitment he had made the year before to broadcast entertainment programs dealing with the population issue. Poindexter then told Stanton what he had heard from Lear about the plan to shelve the controversial episodes. "Are you sure?" asked Stanton. "Yes," Poindexter replied. At that point, Stanton sat back in his chair, smoked his pipe for a minute as he thought, then leaned forward and said to Poindexter: "If I were you, I wouldn't worry about it."[30]

With pro-choice advocates assured, CBS executives chose not to let the rest of the world know about its decision right away. The best strategy was to keep quiet about it until as late as possible, and then quickly slip the two episodes into the summer schedule. The less that opposition groups knew in advance, the more difficult it would be for them to organize another campaign. No doubt Stanton's decision had also been influenced by the Supreme Court's *Roe vs. Wade* ruling. Although at the time of the first broadcasts, a few states—including New York, where Maude was supposed to reside—had legalized abortion, the practice was still against the law in some parts of the country. With the landmark Supreme Court decision on January 22, 1973, abortion became legal in every state. This case clearly gave powerful ammunition to the pro-choice groups and made it easier for CBS to consider rerunning the episodes.

Further encouragement came with the FCC's decision on the Fairness Doctrine case filed against the CBS station in New York. On June 19, the Commission ruled against the complaint by anti-abortion groups, thus assuring the network that it could rebroadcast the programs in question without fear of further legal hassles.

The FCC based its denial on the same grounds that it denied most Fairness cases involving public affairs programming: the failure of complainants to prove that the CBS station had not presented the pro-life side of the issue in any of its broadcasts. In the past the FCC had managed to avoid the question of whether the Fairness Doctrine applied to entertainment programming by ruling that the handling of issues in specific programs was not considered a discussion. But the provocative treatment of abortion in *Maude* clearly amounted to a discussion, requiring the commission to decide the question they had so long avoided. Though the decision made it clear that CBS had no obligation to present both sides of an issue in one entertainment show, it made clear that the Fairness Doctrine did apply to entertainment programming just as it applied to other kinds of programming. We will see later how this decision and others influenced the way networks treated controversial issues in prime time.[31]

Norman Lear anticipated a strong reaction from anti-abortion groups when the announcement was made that "Maude's Dilemma" and "Walter's Dilemma" would be rebroadcast. Between the time of the first protest and the rerun, the producer had hired a specialist to deal directly with advocacy groups. A former aerospace executive and a strong feminist, Virginia Carter became Lear's special assistant during the summer of 1973. Handling the controversy over Maude was her first major assignment. Carter helped orchestrate a campaign in support of the rebroadcast.[32]

In early August, with the reruns scheduled for August 14 and 21, a coalition of pro-choice groups—including the National Organization for Women and the National Association for the Repeal of Abortion laws (NARAL)—began a pre-emptive letter-writing campaign to CBS affiliates, urging them to carry the two episodes. The American Civil Liberties Union was also enlisted for this renewed effort. Telegrams were sent to its chapters around the country calling for members to put pressure on their local stations. The ACLU had stepped in to support television in other cases. Its endorsement of the *Maude* show was an important tactic to broaden the constituency beyond those groups directly associated with the abortion issue.[33]

While pro-choice groups were armed and ready for battle, anti-abortion organizations were again caught off-guard by the announcement of the summer reruns. Bishop James S. Rausch, general secretary of the United States Catholic Conference, accused CBS of "irresponsible and gratuitous" action in repeating the episodes. According to Rausch, Catholic Conference officials had met with CBS president Robert Wood in November 1972, and Wood had given them "reason to believe that the error would not be repeated." CBS denied that such a promise had been made. Outraged Catholic leaders then sent a telegram to the network charging it with "undeniable malice and a calculated intent to offend the sensibilities and deeply held beliefs of a substantial portion of the American public." With only a few days left before the first episode was scheduled for broadcast, the United States Catholic Conference hastily dispatched instructions to the parish priests in its 163 dioceses to campaign against the *Maude* episodes in their next Sunday sermons.[34]

Though only a few affiliates had dropped out of the initial broadcasts the preceding fall, by summer, the *Maude* episodes had generated so much controversy that many local stations in the CBS line-up were squeamish about carrying them. Since both the pro-choice and the anti-abortion forces were engaged in intense grassroots campaigns, the local broadcasters were

Protesters picketing CBS headquarters in 1973 over the abortion episodes of *Maude*. *(Courtesy of the New York Post)*

feeling considerable heat. They were no doubt also disturbed by the FCC's recent Fairness decision. Though it had let WCBS, the CBS-owned New York station off the hook, how could individual stations be insured against a similar costly challenge? It was very likely that anti-abortion groups had been monitoring local programming in the last nine months, gathering evidence for further Fairness cases.

By August 13, the day before the first broadcast was scheduled, 21 of the 198 affiliates in the CBS network had defected. These included not only stations in rural areas but some of those in the big city markets. Coincidentally, the postponed broadcast of the controversial *Sticks and Bones* was scheduled for only a few days later, on August 17, and half of the stations in the lineup were not carrying that program. Many stations were therefore faced with the rare dilemma of having to make not one, but two critical decisions in that week. S. James Coppersmith, vice president and general manager of Boston station WNAC-TV, chose to run *Sticks and Bones* but not to rebroadcast the *Maude* episodes. He justified his decision with the following statement to the press:

> [Sticks and Bones] is contemporary theatre, and I think viewers have every right to see it, experience it and make up their own minds. It is not for the young and immature. We will run cautionary announcements to that effect at the beginning and first commercial break. But, to black out the network on this would be to do a grave injustice to the author, producer and most importantly the adult viewer. This play has something to say, it provokes serious thought, it is an answer to critics of television who accuse television of being cotton candy."

The *Maude* show, he added, was a different matter:

> For one thing, the station has already telecast these *Maude* episodes. Anyone who was a regular fan of *Maude* has already seen them and won't be slighted by our not including them. . . . In addition, there is nothing particularly funny about a 47-year-old woman getting an abortion.[35]

The fact that the station had received hundreds of letters of protest over the *Maude* show was also a critical factor in the management's decision, as it was for every other station head who decided against the reruns. As *Broadcasting* later observed: "The affiliates that refused *Sticks and Bones* did so entirely spontaneously, so far as is known, acting out of genuine dislike for the program and their views of what its reception in their communities would be like. . . . Unlike *Sticks and Bones*," the article continued, "the abortion episodes of *Maude* were the target of an organized campaign."[36]

The pressure campaign had also driven away virtually all of the advertisers. On August 10, Norman Lear informed the press that, while time slots in all of the other *Maude* reruns had been sold out, only thirty seconds of the time available in the two abortion episodes had been spoken for. "This is no mere coincidence," he concluded. Although advertisers offered varying reasons for their failure to participate in the episodes, there is no question that they withdrew in response to the pressure. A General Mills official told the press that his firm stopped all its ads on the series earlier that year for various reasons, but he admitted that the protests after the November episodes—which the company did not sponsor—were definitely a factor. The Pepsi-Cola company, on the other hand, openly attributed its withdrawal to the protests. As the company's spokesman explained, "Our policy is essentially hands-off in terms of any sort of prescreening or prejudgment. . . . [B]ut after the protests . . . the company felt it'd be best to bow out of the repeats rather than antagonize all those people all over again." "You can't win," lamented another advertiser. "We've been getting it from both sides."[37]

Another reason for the advertisers' withdrawal was that, with the show blacked out in so many markets due to affiliate defections, the audience was severely reduced in size. For every station that did not carry the episodes, thousands of viewers were lost. The combination of outside pressures and audience reduction had transformed the *Maude* episodes into a losing proposition, as far as advertisers were concerned.[38]

The two abortion episodes of *Maude* were rerun on the CBS network as scheduled, August 14 and 21. Each was preceded by an advisory which said: "Tonight's episode of Maude was originally broadcast in November of 1972. Since it deals with

Maude's dilemma as she contemplates the possibility of abortion, you may wish to refrain from watching it, if you believe the broadcast may disturb you or others in your family." The cost to CBS for rerunning the show was substantial. No commercials appeared in either of the episodes, and 39 out of the 198 affiliates refused to carry the programs.[39]

But even after the final broadcasts, the battle still wasn't over. Pro-choice groups, angered at the number of advertiser withdrawals, staged a picket at the New York City headquarters of American Home Products, and announced plans for a boycott against the products of the seven corporations which withdrew sponsorship of the two programs: American Home Products, Pepsi-Cola, Alberto Culver, General Mills, Mennen, J. B. Williams, and Pharmacraft.[40]

Other public interest groups joined in the campaign that followed the airing. The ACLU, along with four other groups—the Freedom to Read Committee of the Association of American Publishers, the National Council of Churches, the Union of American Hebrew Congregations, and the Young Women's Christian Association—issued a public statement, congratulating CBS and the affiliate stations that carried the show "for their courageous decision to proceed with the re-run of "Maude's Dilemma" on August 15 and 21, despite pressure to withdraw it from groups who opposed the abortion theme of the two-part episode." Mailed to more than 2,000 television stations, advertisers and public interest organizations, the statement warned:

> The public interest is not served when a station's program decision is made on the basis of fear of controversy, or when an advertiser's sponsorship is determined by fear of economic reprisal. Such action by stations and sponsors could become an extremely dangerous precedent for the future of commercial broadcasting in this country.[41]

The National Citizens Committee for Broadcasting announced that it was going to ask the FCC to investigate the stations that refused to carry both *Sticks and Bones* and the two *Maude* episodes. Such action was a violation of the Fairness

Doctrine, argued the media reform group, because the regulation required not only that both sides of controversial issues be broadcast but that stations air controversial material in the first place. NCCB also asked the FCC "to insert a copy of its complaint in the renewal file of every affiliate that turned down the programs, to count against the licensee at renewal time if such rejections reflected station policy concerning the airing of controversial programming."[42]

The debate over *Maude*—and pressure groups in general—continued in the press. A *New York Times* editorial, reviewing the controversies that year over CBS programs, stated: "The public has a choice of liking or disliking these or any other shows, from dramas to documentaries. But if the networks must bow to every offended organization or worry about offering response time for every adult theme, blandless [sic] will become even more totally the norm than it already is in a medium governed by the tyranny of ratings. The right to watch, to hear, and to read cannot be defined and delimited by pressure groups in a mature society."[43]

But the Catholic press raised other issues about the controversy. Robert B. Beusse and Russell Shaw, in an article titled "Maude's Abortion: Spontaneous or Induced?," questioned whether "a hard-driving pressure group called the Population Institute" should have that much influence on prime-time programming. The *Maude* episodes, commented the article, "raise the fundamental issue of who—in this instance at least—was calling the shots on what Americans were to see and hear concerning the controversial issues of abortion and vasectomy on CBS television." "There is something strange," the writers suggested, "in the fact that protests and pressure against the *Maude* shows on abortion evoke outcries in some quarters; while the pressure that helped bring Maude's pro-abortion decision into millions of living rooms is passed over in silence." Finally, the article concluded, "perhaps the strongest lesson in all this for pro-life people is to 'go and do likewise.' Protest after the fact is often necessary and sometimes (as in the case of the *Maude* reruns) can produce results. In the long run, however, it is even more important that the pro-life movement take a leaf from the book of groups like the Population Institute and conduct the well-planned, well-organized efforts needed to get a fair hearing in the communications media today."[44]

While the advocacy groups and the public debated the issues, the network television industry, in the aftermath of the *Maude* controversy, had to face its own problems. In this "new era of candor," as *Time* magazine had labeled it, the networks—particularly CBS—had faced a number of organized groups capable of going to great lengths to show their displeasure with television programming. And, as the *Maude* case had demonstrated, when the issue was a controversial one, the networks could be subjected to pressure from groups on either side. Many people were predicting that the impact of all these pressures would be to make television programming bland. But, in this "new era," bland was not the order of the day. Controversy, even with its concomitant disruptions, had great drawing power for American audiences. And though advertisers did not like being caught in the middle of messy situations that might cost them money, they also did not want to risk losing audiences if prime-time programming reverted back to its earlier, pre-*All in the Family* state. The comments of Pepsi-Cola's media director Henry Hayes are particularly noteworthy. In a trade publication article, the executive warned that advertisers should "not try to control programming through economic clout." While appearing to voice First Amendment concerns, Hayes was really more alarmed about threats to business. Should television "lose its vitality" from nervous advertiser pressure, he predicted, the result would be a "lower level of sets-in-use by the public."[45]

For Lear, the dilemma was even more critical. His whole operation was based on controversy. The *Maude* experience had been a sobering lesson for the producer. He told the press a few days after the second broadcast that "the network said to me by telephone: 'Norman, we're hearing from so many groups asking questions. . . . Do you feel you have a right to put so much opinion into a situation comedy? You have people tuning in to see a comedy, and while it is entertaining, you're also delivering a message. Do you have a right to do it?' "[46]

Lear's reply to the query was: "I don't know another way to create theater every week except to deal with those situations I find important in human values." But Lear and his staff had spent an incredible amount of energy fighting the political battles that erupted over two episodes of one of his comedy series. If he were to continue to deal regularly with controversy, he would need to work out his own system for managing the trouble his shows engendered.

CHAPTER FOUR

Managing Advocacy Groups

Justicia was at war with the networks. Shows like *Ironside, Alias Smith and Jones,* and *The Wonderful World of Disney,* charged the group's leaders, "portray Mexicans as overly meek, consider us recent transplants from Mexico, cast us as hat-in-hand types, gives us mascot roles, and generally find us the target of a white hero." To correct these injustices, Justicia ("justice" in Spanish) made several demands. Not only must "demeaning shows" be dropped from the air, the activists insisted, but the networks should also allow Justicia to monitor all scripts and pre-screen all programs with Chicano characters in them. Finally, the advocacy group called for the networks to set aside $10 million to pay for new shows with Chicanos in "significant roles." This money would be a "compensatory fund . . . for demeaning Chicanos and keeping them out of the industry for so long."[1]

Justicia launched its attack during the summer of 1971, knowing that all California TV licenses were up for renewal that fall. If the networks refused to comply with the demands, Justicia was prepared to file petitions with the Federal Communications Commission, asking the agency to take away the broadcast licenses of the network-owned TV stations in Los Angeles.

In the early seventies, the networks were faced with two interrelated trends. While the business of entertainment television was driving its programming in the direction of more socially and politically relevant material, activist groups were making their own demands on prime time. Representing not only Chicanos but women, blacks, Asians, and gays, many of these groups were motivated to political action by their very invisibility in the media. Though they called for the elimination of stereotypes, they also insisted on wider representation. They launched well-orchestrated—sometimes militant—campaigns to force the networks to include them in entertainment programming.

To some degree, these two trends were compatible. In the same way that population control issues could stimulate more interest in prime time, meeting minority group demands could help to make programming more "relevant" and therefore more appealing to the younger, urban audiences that were becoming so desirable to advertisers. But, the more programs incorporated ethnic characters or controversial issues, the more likely that groups with a concern for these issues would be vocal and active. Just as *Amos 'n' Andy* and *The Untouchables* in years before had served to mobilize black and Italian American groups against the networks, so the programming of the 1970s functioned as targets for the proliferating political advocacy groups. And for nearly every issue, there was an advocacy group with a stake in how that issue was treated. As relevancy became an increasingly profitable element of prime-time programming, advocacy groups became an increasing presence and an aggravation.

For the networks, the bottom line was protecting their profits. When a show such as *Maude* was broadcast without any commercials, it resulted in a substantial loss of advertising money to the network. When groups challenged the license of a network-owned station, the legal costs were considerable. The pressures on prime time were a problem not only for networks but in some cases, for advertisers and affiliates as well. Advertisers didn't want the responsibility they once had for program content, but they also didn't want to be made vulnerable by the programs in which their commercials appeared. The last thing advertisers needed was to be the target of an organized boycott. At the same time, advertisers, networks, and affiliates alike looked to the networks to deliver the programs that would be controversial enough to draw large, demographically desirable audiences. But they also looked to the networks to protect them from the pressure that might be generated by those programs. If network television was to continue broadcasting controversial, issue-oriented programming, the industry would need to work out strategies for maintaining the delicate balance between controversial content on the one hand, and smooth relations with advocacy groups on the other.

This task was assigned to the network standards and practices departments. In their expanded role since the quiz scandals, these departments were assuming more and more responsibility for protecting the networks from external criticism. Standards and practices departments performed this function in two ways. As censors, they supervised the writing and production of television programming, making sure that no program element was broadcast which could result in criticism from either the government or the public. In their "public relations" capacity, these departments handled complaints, and in doing so, acted as a buffer for the other departments within the network corporation, as well as the affiliate stations, the production companies, and the advertisers.[2]

Though standards and practices departments had routinely dealt with complaints in the past, their direct involvement with advocacy groups had been infrequent. Not only had there been fewer groups to contend with but content policies had also served to prevent objections. While the NAB Code remained the official public industry policy for television programming content, the actual content policies were the unwritten ones devised for pragmatic reasons by the network standards and practices departments. Often the networks had found it advantageous, for example, to avoid certain issues and obscure the identity of ethnic characters in entertainment shows as a strategy for preventing complaints. As one reporter had noted in 1961:

> There seem to be "built-in" defenses by TV against serious complaints—though they do flare up—by the medium's emphasis on 'non-controversial' programming. Writers usually shy away from combustible themes because they have learned they are not acceptable generally to advertisers and networks and stations. [Standards and practices] departments keep a sharp eye

> peeled for themes or characterizations or dialogue that might prove offensive to a specific group. . . . Partly in jest, one producer outlined the details of an "acceptable" western series: "It would star a white Protestant of no particular denomination; he would have no known occupation and no known place of birth; he would have no visible income. In short, he would be a handsome 'nothing.' Who could complain?"[3]

But, by the 1970s, the combination of increasing controversial content on the one hand and more political activism on the other made the job of the standards and practices departments more challenging. Neither groups nor issues could be dismissed or obscured so easily.

Standards and practices departments also became increasingly more important to the TV networks. With the potential disruption and financial loss that outside pressure posed, the dual functions served by these departments gave them an indispensable role in protecting the economic well-being of the network institutions. More than ever before, it was essential that the networks centralize their control of content and their interaction with organized groups in one department. By doing so, criticism could be channeled into a manageable form. One department could keep track of which groups were most active, and what their reactions had been to content. This same department could then ensure that the programming which was distributed to affiliates across the country and in which advertisers placed their commercials was safe. In this way, standards and practices departments increasingly were not only censors of content but also "sensors" of public taste and acceptability.

When many of the more militant advocacy groups first began to take aim at prime time, standards and practices executives hastily devised stopgap, defensive measures to keep these groups under control. Before long, these ad hoc tactics evolved into a more coherent strategy for dealing with advocacy groups. Within a very few years, the network industry had shifted its overall strategy for handling these pressures from a reactive to a proactive one, developing a full-blown system for "managing" advocacy groups. As this system became more institutionalized, it served to disarm, contain, and control most of the groups. The groups in turn, had to learn to adapt their strategies to fit the network system. Those groups that were unable to adapt were generally replaced by more moderate organizations whose leaders were willing and able to work within the system. Though there were periodic "breakdowns" where miscalculations caused disruptions, these experiences generally served to fine-tune the system. They became valuable lessons for the networks, functioning to make the system an even better mechanism for control of outside pressure.[4]

Justicia's involvement with entertainment television took place during the initial period when the network "containment" strategy was being developed. Though the group vanished within a few years of its first appearance, it was not forgotten by industry executives. The story of Justicia's short-lived, but memorable, sojourn into the backstage world of prime-time television sheds light on the dynamics of this early period of confrontation between political advocacy groups and the networks.

Justicia was only one of a number of national Mexican American organizations active in media reform campaigns during the late sixties and early seventies. Much of the work of groups such as the National Latino Media Coalition, La Raza, the League of United Latino Citizens (LULAC), and the Mexican-American Anti-Defamation Committee focused on grassroots efforts for reforming local television, although they made some moves to change entertainment television. Two well-known television characters disappeared from the airwaves because of pressures from Mexican American groups. The Frito-Lay Company found itself under attack because of its "Frito Bandito," who galloped in and out of commercials for Fritos Corn Chips, absconding with the popular snack food. The Mexican-American Anti-Defamation Committee charged that the famous cartoon character carried a "racist message" that Mexicans are "sneaky thieves." Within a short time, the Frito Bandito had galloped off the screen altogether. Mexican American groups also successfully purged another familiar Hispanic character from prime time. "Jose Jimenez," reported the *New York Post* in 1970, "the zany, not-so-bright, happy-go-lucky character created by comedian Bill Dana, is dead—by request of some Mexican-

Americans." The Congress of Mexican American Unity held a mock funeral for Jimenez in 1970, where Dana himself read the obituary. The comedian had made a successful television career by portraying the dull-witted character whose famous line, "My name, Jose Jimenez," elicited laughter on countless variety shows. But Mexican American activists had complained that the character was racist, and Dana voluntarily agreed to kill off Jose, even though, as he told reporters later, he didn't personally feel the character was damaging to Mexican Americans.[5]

While these other national groups made sporadic attempts to change prime-time TV programming, Justicia was in a unique position to launch a more direct campaign targeted against entertainment television. A small, Los Angeles-based community group, Justicia, had broken off from another local organization called the Mexican American Political Association (MAPA) in 1969. MAPA had begun a campaign against the film and television industries, threatening to boycott theaters unless studios changed their hiring practices and portrayals of Mexican Americans. Justicia decided to focus most of its efforts on prime-time television. Though the group was small and its resources scanty, Justicia's membership and its location gave it a critical advantage over some of the larger, national Mexican American organizations. Its tiny storefront headquarters in East L.A. were only a few miles from major production studios and the West Coast offices of the three networks. Justicia's members were young, energetic, committed activists. Some were students at nearby California State University, Los Angeles; a number of them were professional actors who knew something about the entertainment business. Justicia's president was Ray Andrade, a spirited young man of twenty-five, with the diverse background of having grown up on the streets of East L.A., served as a Green Beret in Vietnam, boxed professionally, and acted in motion pictures.[6]

Justicia members were in communication with other activist groups at the time, including the Congress on Racial Equality (CORE), which was engaged in its own pressure campaign against the networks. Both groups were part of a network of organizations involved in civil rights activities nationally and locally. Some of these groups held regular meetings in people's

Chicano protesters demanding representation in film and television during the early seventies. *(Courtesy of Wide World Photos)*

homes, and it was at one of these meetings that Andrade remembers encountering political organizer Saul Alinsky. Alinsky tutored Andrade in the strategy of activist leadership. With his skill as an actor and his rather flamboyant personal style, Andrade already had the qualities for charismatic leadership. Andrade recalls, Alinsky's advice: "You've got to be a little crazy," he explained to Andrade, "and you've got to show that you have a sense of humor but a serious side at the same time."[7]

The group needed an effective method for pressuring the media. For this, they went to industry insiders, to the university, and to other groups in the media reform movement. In particular, Andrade remembers that Albert Kramer, of the Washington-based Citizens Communication Center, instructed him in some of the recent developments in FCC regulations, particularly the use of the petition to deny. Justicia decided to join other groups in a coordinated attack on the stations in the Los Angeles area. Justicia members really weren't interested in taking the broadcast licenses away from the network-owned stations. The group used the threat of a license challenge to gain the attention and cooperation of the networks. The fact that other groups were actually proceeding with formal license challenges against some of the seven stations in the L.A. market helped to create a general environment which encouraged cooperation from industry executives.[8]

Challenges against network-owned stations posed very real economic threats. Tom Kersey, ABC's West Coast vice president for broadcast standards and practices, recalls that the corporate leadership at his network was very concerned about the potential cost and disruption of a license challenge and willing to do a lot to avoid it: "Any challenge to a station would be an enormous threat. One of our stations was worth between $35 and $40 million. We pulled the legal brains together and [decided that] the events, the whole routine would be expensive beyond belief."[9]

ABC's first move in its dealing with Justicia was to arrange a meeting between the group's leaders, local station management, and top-level executives from the network—including its president, Elton Rule, who flew out from New York to the KABC station offices in Los Angeles. Predictably, this initial encounter between the corporate executives and the streetwise activists was less than cordial. Participants from both sides disagree on exactly what happened during the meeting. As Kersey remembers, the Justicia leaders threatened more than legal action: "The challenge was that if we didn't sit up and listen and in fact execute all of the demands of Justicia, they would have Brown Berets up there with guns and force us to meet their demands." Ray Andrade, on the other hand, says that Kersey exaggerated; Justicia never threatened any violence. But, regardless of whose memory is accurate, what is important is the extent of the threat the network executives perceived and their willingness to cooperate with the pressure group. "It was Elton Rule's position," recalls Kersey, "that they [Justicia] had some legitimate concerns and we had some legitimate responsibilities."[10]

To pre-empt any further confrontations with the activists, ABC management decided to solicit input from the Hispanic group members on certain upcoming programs. As Kersey recalls, "We agreed to take on Paul Macias and Ray in a consultant capacity, so every time we had a Mexican-American family or a [character], we would talk with them [Justicia] to get their responses."[11]

The practice of using "technical consultants" was already commonplace in the television industry. Generally, it had been restricted to organizations with highly specialized expertise in a given area, though the lines between technical assistance and advocacy were not always that clear-cut. In the *Maude* case, for example, the same organization that provided script consultation had also orchestrated a supportive campaign for the abortion/vasectomy episodes. The more typical use of technical consultants was in TV programs involving specialized professions like law, medicine, or psychology. Networks and producers routinely used experts from the medical field, for example, to ensure accuracy in series such as *Dr. Kildare* and *Marcus Welby, M.D.* Such cooperative arrangements had proven mutually beneficial: the networks used the outside groups for ideas and as a hedge against potential complaints, and the trade associations used the network programs as public relations vehicles. In the mid-sixties, for example, NBC and MGM had worked with the National Education Association on the popular series *Mr. Novak,* which starred James Franciscus as a young teacher. The NEA had set up a rotating panel of teachers to serve as ongoing consultants and to review all the scripts for the series. Panel members suggested themes for the show and provided input as to how those themes should be handled. The NEA saw *Mr. Novak* as an effective way to enhance the image of teachers, and they tried their best to make that image as positive as possible.[12]

By the 1970s, with groups like Justicia placing more pressure on the networks, the practice of technical consultation was now being adapted as a political tactic. Standards and practices executives at the other two networks also worked out arrangements to give Justicia input on prime-time programs. Tom Downer, CBS program practices director, told *Broadcasting* magazine that Justicia had pre-broadcast access to his network's programs. Though the executive claimed the group had no veto power over either shows or commercials, he defended the practice of using Justicia as a consultant. "What offends Chicanos is changing constantly," explained Downer, "and that's why it's valuable to meet with Justicia. They definitely perform a useful service as far as we're concerned." Jack Petry, NBC's West Coast standards and practices head, told the press that his "dialogue" with Justicia had been productive. The group had come to him with some complaints and he had tried to satisfy them. In his opinion, the executive added, NBC was doing "quite a bit to aid the Justicia cause." Standards and practices departments during this period would routinely refer scripts to Justicia for input. Justicia had a "monitoring committee" charged with the responsibility of reading the scripts, and Andrade and Macias would report the committee's comments back to the network. Standards and practices executives would then negotiate with producers for the changes.[13]

Though the decision to grant Justicia script review privileges may have served initially to disarm the advocacy group, the practice ultimately failed. For the networks, giving such a militant group early access to programs proved to be disastrous. There were problems with it from the beginning, recalls Kersey. "The first show we pulled them in on a consultancy matter, he [Andrade] threatened to have the studio shut down. . . . Oh, there was so much emotion . . . we didn't know how to deal with it really." Nor did the procedure work for Justicia. The combative style of the group's leaders that had earned them entry to network decision makers failed to serve them well on an ongoing basis.[14]

One of the more dramatic incidents involved an NBC Western series called *Nichols*. Justicia had heard of the series while it was still in development. The activists were angered because no Chicano appeared in the program, even though the story was set in the American Southwest at the turn of the century, when Mexican Americans made up a large segment of the population in that region. Justicia demanded that NBC incorporate a Chicano as an ongoing lead character into the show and that the network "guarantee that all segments of the show properly portray the Chicano with dignity and that the Mexican-Americans be shown as contributing to the history of the Southwest." The show's producer, Meta Rosenberg, didn't see it that way. To Rosenberg, Justicia's demands were unreasonable. Andrade not only wanted a regular Mexican character, she remembers, "he wanted something that was really out of the question; he wanted a Mexican producer or writer on the staff. Well, he has no authority and no right to ask us for that. That is ridiculous."[15]

As a producer, Rosenberg's position on the issue was fairly typical. The networks were feeling the threat of the pressure groups more than the producers were, since it was the network-owned stations whose licenses were vulnerable. Studios and production companies had less to lose from pressure groups than the networks. Many of them found the network insistence that they comply with the demands of these groups to be yet another intrusion into their creative freedom. Rosenberg's power relationship with the network on this series, however, was not typical. Her exclusive contract for the show's leading man, James Garner, gave her considerably more leverage.

Since Rosenberg appeared unwilling to make any changes, the standards and practices executives suggested that Macias and Andrade meet directly with her. Minutes after the meeting began, Andrade leaped onto a chair and began screaming at the producer, who immediately ordered him out of her office. The Justicia team returned to NBC, where they were told that there was little the network could do, since the producer could decide to take the entire series away.

The Justicia leaders then decided to wait until the show aired, find out who the sponsors were, and go after them. One of the show's sponsors was Chevrolet. The activists wrote to the company with their complaint, but got no cooperation. So they decided to organize a boycott. "The idea of a boycott seemed like a good one to us at the time," recalls Andrade, "since the Cesar Chavez boycott against California growers had been successful." Justicia leaders did not have the resources nor the constituency to launch a national boycott campaign, but they could take action in their own neighborhood. "A lot of Mexicans were driving Chevys at that time," Andrade explains, so the group organized a picket against three of the Chevrolet dealerships in East Los Angeles. The terrified manager of one of them immediately called General Motors headquarters, and GM executives frantically phoned the network. Within a few days top network executives had flown out from New York to meet with Justicia at the group's storefront headquarters. Although this dramatic event did not produce any concrete changes in the show, it left an impression on NBC executives. *Nichols* survived one season on the network, and was canceled in August of 1972, due to low ratings.[16]

Justicia itself had disappeared from the scene by 1972, when its leadership fell apart. Though Justicia's involvement with network television had been brief, its success in gaining access to high-level decision makers had been remarkable. The activist group might have been able to continue working with the networks if it had altered its style. But the explosive, unpredictable behavior of the group's leaders ultimately undermined their effectiveness. As other advocacy groups were learning, dealing successfully with network television required cooperative, as well as confrontational strategies.[17]

Though Justicia had vanished, Ray Andrade surfaced a few years later when producer James Komack hired him to work on *Chico and the Man,* a new comedy series about an older white man and a young Chicano boy. Komack met Andrade during

Justicia's heyday. In fact, the series was loosely based on the activist's own experiences growing up in East Los Angeles. Though officially hired as the associate producer, Andrade's role was also intended to pre-empt attacks from the Chicano community. The series was first proposed to ABC. After that network turned it down, Komack took it to NBC, where executives liked the proposal but told the producer they were afraid to take it because they didn't know how Justicia might react. Though the group had disappeared, the network didn't know when or where it might reappear, and executives hadn't forgotten the trouble Justicia had given them just a few years earlier. Komack assured executives that Justicia would not be a problem because Andrade was working for him.[18]

But if television executives thought that hiring Andrade would prevent complaints from Chicanos, they soon found that they were wrong. Shortly after the series premiered, a small group of Mexican American activists, some of whom were former members of Justicia, launched a protest, complaining that the series did not reflect Mexican Americans accurately or fairly. They were particularly unhappy that the show's lead character was played by a Puerto Rican, Freddie Prinze, and not a Chicano. Though Andrade was working for the producer, he decided to join the picket outside of the network headquarters. As he explained it, "there was such a small group of people out there, I thought they needed some help." Andrade's remarks to the press revealed the complexity and ambiguity of his role in the series. It's clear that he really wasn't sure where he stood or what he was supposed to be doing. In an interview with *Daily Variety,* he attacked the show, charging that "the character of Chico is cheap and demeaning. The old man at one point tells the kid to get out of the garage and take your flies with you. Chicanos didn't like that line." But in the same interview, he added: "I think the show has great potential, and I am 70% happy with it, but there is a lot to explore in terms of authenticity." Asked about his role in the production process, he answered: "I have nothing to do with the creative part of it. They asked me at first, but didn't like my suggestions and ignored me. . . . I tried to sell my ideas to the producers, but it was zilch—I got nowhere. They are suspicious of me, they don't know if I'm going to undermine the show. They go to their Beverly Hills homes and I go to mine in East L.A., and there is suspicion on both sides." Andrade finally left the show voluntarily, although with the encouragement of the producer, who could see that having the former activist on the staff was causing more problems than it was solving. Though some Chicano groups continued to protest it, *Chico and the Man* remained on NBC until 1978.[19]

Executives at all three networks were no doubt relieved with the departure of Justicia and its leaders. But they were still faced with periodic complaints from various Hispanic groups. Rather than deal directly with all of these organizations, standards and practices executives found it advantageous to cultivate a relationship with one group. Recognizing a single organization as representative of all Hispanics not only simplifed network relationships with advocacy groups but also created a shield against criticism.

ABC's Tom Kersey, who spent more than twenty years negotiating with advocacy groups, labeled this policy the "one voice concept." In time, it became a strategy adopted not only by the networks but also by a number of the advocacy groups themselves, who found it more effective if one organization was designated as the official group to deal with the networks on behalf of a particular constituency. Explained Kersey: "The community speaks with one voice. That is what happened with most of the groups. They have organized themselves. . . . When they did that and their one voice could be heard, then we paid more attention, we listened more carefully, we got more out of what their concerns were and we were able to address them."[20]

More often than not, it was the moderate groups, those that showed they were willing to play by the rules, that were granted the "one voice" status, ultimately gaining access and cooperation from the networks.

It was on this basis that the networks sought out Nosotros. Founded by actor Ricardo Montalban in 1970, and based in Los Angeles, Nosotros served both as an advocacy group and a casting agency for Hispanic actors. This dual purpose meant that Nosotros maintained a cooperative stance toward network television. As explained in one of the group's pamphlets: "We seek only to work within the system, with the abilities we have, to improve the image and ambitions of those 16 million persons of Spanish-speaking origin in the United States." While Justicia members picketed the Oscar and Emmy awards ceremonies for not paying enough attention to Hispanics, Nosotros began giving out its own "Golden Eagle" awards to honor Hispanic actors.[21]

Network standards and practices executives found this moderate approach much more agreeable than what they had experienced with Justicia. And by the mid-seventies the networks had begun to work regularly with Nosotros representatives, invoking their relationship with the group whenever complaints or questions came up about Hispanic images in television. As they had done with Justicia, the networks periodically referred scripts to Nosotros representatives for commentary and advice. Unlike Justicia, however, Nosotros members were a good deal more willing to compromise. Although there were occasional protests from the group, for the most part it maintained friendly relations with the networks. The dependency of the Nosotros members upon the television industry for their livelihood made it difficult for the organization to engage in militant behavior.[22]

The "one voice concept" was only part of what evolved into a general system for managing advocacy groups. After their initial period of struggle and experimentation with the new breed of advocacy groups in the early 1970s, standards and practices departments soon developed routines for dealing with these organizations on a regular basis. Though each network corporation developed its routines separately from the others, the general policy followed by all three networks evolved along similar lines. This is not surprising, since all three networks faced essentially the same groups and had the same functions to perform within their respective companies.[23]

Open-Door Policy

Even though the standards and practices departments preferred to deal with more moderate advocacy groups, they adopted an open-door policy for "special interest groups"—the term the networks applied to all groups trying to influence television. This policy had certain strategic advantages. Since many of the militant groups felt shut out by network television, merely allowing them to enter the corporate offices of the major networks was one effective step toward disarming them. Some of these first meetings were anything but friendly. Frequently they erupted into shouting matches. But the network executives proved skillful at turning what often began as hostile confrontations into disarming "dialogue" sessions. Routines were developed for handling these meetings. Guests would be invited into plush offices in New York or Los Angeles, greeted by a team of network executives from various departments, and encouraged to articulate their concerns and complaints about prime-time TV.

These meetings were also used to educate outsiders about the television business, particularly the constraints under which network television operates. While promising to do their best to correct injustices and improve treatment of advocacy group issues, network executives would de-emphasize their own control over programming, shifting responsibility to other less accessible portions of the industry. As explained by one New York-based standards and practices executive: "We tell them [the group representatives] that essentially we buy programming and that we buy from the production community [in Hollywood] and while we do have some control over making sure that it meets our standard, we don't write it. We don't initiate it." Reported one group leader: "We heard this one at all three networks: 'We're not responsible. That comes from the creative community and that's out on the West Coast." Though this explanation may have been an effective political tactic, it was not an entirely accurate description of the network role in production. While most prime-time programs were produced by production companies in Hollywood, the networks were involved in a great deal more than merely "buying" the shows. They partially underwrote the productions, and they decided what would be produced, what would be broadcast, and what would be canceled. Though these facts were commonly understood by anyone familiar with the television industry, they were not so obvious to outsiders.[24]

In addition to indoctrinating advocacy groups, meetings served an important safety valve function by allowing group leaders to articulate their concerns and vent their anger. Such face-to-face encounters also had a way of humanizing the networks and making them appear accessible and accommodating. Standards and practices executives, skilled in interpersonal communication, routinely made promises to be "sensitive" about the group's issues, and to continue an "ongoing dialogue" with them. Noted one advocacy group leader, "It is hard to hate people you've met."

Ongoing Relationships

Standards and practices executives generally kept their promise to remain in contact with advocacy groups. Like the open-door policy, the policy of developing "ongoing relationships" with such groups also served strategic purposes. Maintaining frequent contact with group leaders made it possible to keep the lines of communication open, consequently reducing the possibility of an unexpected protest. Certain executives within the standards and practices departments were assigned the responsibility for maintaining such ongoing relationships. One of these was NBC's Bettye Hoffman. From her New York office in Rockefeller Plaza, the executive became involved not only in coordinating and following up the meetings with advocacy groups but also in "seeking out" and initiating contact with them. Sometimes this would mean traveling to the groups' headquarters in various parts of the country. As she explained it:

> I try to get together with as many of them as I can at least a couple times a year. And when we have our annual convention in California . . . I generally tack on two to three days in order to make contacts with groups out there. People in Washington are closest so that I make special trips down there from time to time to drop in on them, talk with them and find out what problems they have.[25]

This practice enabled the networks to keep track of the activities of political advocacy groups and to anticipate any pressures that might be directed toward the television industry.

The groups that were able to sustain the longest ongoing relationships with network television did so by adapting their strategies so that they were compatible with the policies and procedures of the standards and practices departments. These groups also tended to be connected with "manageable" issues, whose treatment in entertainment programming could be negotiated without major disagreements.

Lydia Bragger, chairperson of the New York-based Gray Panthers Media Watch Task Force, first approached the networks with a complaint. As chief media spokesperson for older Americans, Bragger wrote a strong letter in the mid-seventies to the president of CBS, charging that a segment of the *Carol Burnett Show* had portrayed the elderly as "cranky, stubborn, crumbling human beings." Following its usual policy of channeling such complaints, the network passed the letter along to the standards and practices department. Executives there quickly telephoned Bragger and arranged a meeting.[26]

Bragger's own background as a media professional—she had worked in public relations and broadcasting for years—equipped her with an understanding of the way the television industry operated. It also made her sympathetic to the network point of view and willing to compromise. In Bragger's words, "I never antagonize them; I approach them on the premise that they're not aware of what they're doing." Within a short period of time, Bragger was on a first-name basis with the executives in the New York standards and practices offices at all three networks. Relationships were cordial and routine. If there was a problem with any portrayal in prime time, explained one executive, "Lydia would come in and we would sit, some three or four of us, from the network. We would look at this segment with her (Lydia) and then we would discuss it. If there was cause for concern, we would let our West Coast broadcast standards people know that Lydia was in and that we had looked at this segment, that it could have been done better, that certain elements might have been changed to avoid the problem in the future."[27]

The Gray Panthers became very helpful to the networks, offering commentary and critique on completed programs, and input and advice on future ones. Standards and practices executives became sensitized to the issues that Bragger and her organization were concerned with. Most of the adjustments that the Gray Panthers wanted in programming could be quite easily accommodated. "They (the Gray Panthers) don't like 'senior citizen,' " NBC's Hoffman noted, "and I think what we settled on was the 'older adult.' "[28]

Sometimes the Gray Panthers were more extensively involved in consultation on the writing and production of specific shows. When producers of the CBS series *Lou Grant* decided to do an episode on the abuses of the older Americans in nursing homes, the network referred them to Bragger and her associates. Using the advocacy group representatives as technical consultants, the producers solicited input from the Gray Panthers throughout the writing process. After the broadcast, the episode drew fire from nursing home trade associations, whose members charged that it unfairly portrayed their profession. When pressure on sponsors threatened to keep the show from airing a second time later that season, the Gray Panthers launched a counter-campaign to persuade advertisers to stay with the controversial segment, much as the pro-choice advocates had done a few years earlier during the *Maude* controversy. For CBS's stalwart commitment to carry the show in the face of sponsor withdrawal, the Gray Panthers subsequently gave the network an award.[29]

Input

Soliciting input from advocacy groups on upcoming network programs was yet another practice that had become institutionalized by the mid-seventies. The degree of advocacy group involvement in the creative process could vary substantially, depending on who the group was, and the nature of the project. Standards and practices editors might flag a line of dialogue or a plot sequence that could be offensive to an advocacy group, and then "run it by" a group representative over the telephone. Or the standards and practices department might ask an advocacy group to read portions of the script, advising on how sensitive issues might best be handled to avoid controversy. From time to time, completed programs would be pre-screened to advocacy groups prior to broadcast.

These practices were handled carefully and with as little public knowledge as possible. The experience with Justicia had taught the standards and practices departments a valuable lesson. As a consequence, not all groups were afforded such easy access to pre-broadcast material; those that were often agreed to keep quiet about it. Sometimes even the producers were not told of this consultation with advocacy groups. If a program required extensive input from an advocacy group, standards and practices executives would often insist or strongly suggest that a producer "work with" the group during the writing and production process. Generally this consultation was not indicated in the credits, although occasionally the assistance of an expert might appear.[30]

Once in a while the practice backfired. After his success with *Roots,* the ABC miniseries that traced several generations of black families, producer David Wolper decided to adapt Ruth Beebe Hill's book, *Hanta Yo,* along similar lines. ABC standards and practices insisted that Wolper seek out representatives from the Native American community to consult with them on the program. Wolper did so, but found that, when he began to meet with the consultants, they were so unhapy with the book in the first place—which they saw as inaccurate and exploitative—that they believed nothing could be done to salvage it. Group members went to the press with their complaints, rallying support from other Native American groups and stirring up such a controversy that the show was shelved for over a year. After the trouble died down, the miniseries was quietly picked up again, a "moderate" Sioux consultant—described by *TV Guide* as "the agreeable malleable editor of the *Lakota Times*"—was hired to complete work on it, and *Hanta Yo* was renamed *The Mystic Warrior* and broadcast without any major public protest.[31]

Assimilation

Such costly and disruptive experiences were a continual reminder to standards and practices departments that the process of soliciting input from outsiders could be risky. A more reliable method of securing input was to hire people in the standards and practices departments who could represent advocacy group interests. This policy could reduce the need to rely on outside consultants, while at the same time offsetting criticism. Hiring minorities and women into standards and practices departments was also a manageable method of meeting the demands by advocacy groups for increased representation within the network corporations. Once they came to work for the networks, these employees would learn to adapt to the corporate world and to adopt a "mind set" compatible with industry constraints.[32]

By the mid-seventies, the networks were publicly characterizing their standards and practices departments as microcosms of pluralistic American society, staffed not only with blacks, Hispanics, women, and Asians but also with senior citizens, Catholics, and gays. Several networks employed members of advocacy organizations such as the NAACP and Nosotros.

A parallel system for managing advocacy groups was developed by Tandem Productions, the production company run by Norman Lear and his partner, Bud Yorkin. When Lear hired Virginia Carter—in the middle of the *Maude* controversy—he assigned her the responsibility for negotiating directly with advocacy groups. This move set Lear apart from other Hollywood TV producers, who preferred to let the networks handle the public. Lear needed his own political specialist to prevent the kind of disruption he had experienced with the *Maude* episodes because, in Carter's words, "controversy is our bread and butter." In all of the shows Lear had running on the networks in the mid-seventies—which included *Maude, Good Times, Sanford and Son, The Jeffersons, One Day at a Time,* and *Hot l Baltimore*—racial issues, religion, homosexuality, and sex were staples. The producer's decision to deal directly with advocacy groups was a key to his success in television: it gave him more autonomy from the network standards and practices departments; it defused a good deal of the pressure generated by his controversial shows; and it gave him direct access to a myriad of groups that could provide material for his programs.[33]

Carter functioned as a mini-standards and practices department for the production company. Like her network counterparts, she developed policies for disarming and containing advocacy groups. She developed an open-door policy, met frequently with various organizations, solicited input for scripts, pre-screened programs, and established ongoing relationships with a number of advocacy groups. Like the network strategies, these procedures served the dual function of providing producers with useful information as well as preventing protests. Carter also pioneered the idea of using her contacts with these organizations to promote the shows among grassroots organizations around the country, a practice that later became commonplace with the networks.[34]

Though generally effective, Lear's strategies didn't entirely free him from attack from advocacy groups. Black groups took the producer to task in the mid-seventies for his comedies *Good Times, Sanford and Son, The Jeffersons,* and *That's My Mama.* These shows, argued Pluria Marshall, head of the National Black Media Coalition, "are about blacks but not for blacks. . . . [They] are produced by white folks for white folks. . . . [T]he men always have trouble finding jobs or keeping jobs[,] . . . not a very positive image for my son to watch and try to emulate." Charles Cook, of CORE, after meeting with Tandem to complain about these shows, told the press, "We get coffee from Virginia Carter, not solutions." When actor John Amos left his role as the father on *Good Times* in 1976, black groups complained that without Amos, the show would reinforce the stereotype of black matriarchy. But, while the *Good Times* incident angered black groups, Lear's other policies toward blacks offset some of the criticisms. He not only employed a number of black actors in his comedies but also instituted a black writers' program at Tandem. He was a strong advocate for racial justice, often donating considerable money to the cause.[35]

The networks adapted quite effectively to the pressures they faced from advocacy groups in the early 1970s. Though the legal powers that advocacy groups had acquired in the sixties had successfully gained them access to network corporate headquarters, the networks had found ways to set the terms for continuing access. The system for managing these groups would continue to evolve. It was by no means perfect; breakdowns—sometimes of large proportions—would occur. But the essential mechanisms of control had been tested and refined and found to function adequately at protecting network television from major threats to business. The advocacy groups trying to influence prime-time content in the seventies and eighties would encounter a team of experts within the networks, equipped with a sophisticated set of skills and strategies designed to minimize disruption and maximize friendly, cooperative relationships. Successful advocacy strategies would have to take all of this into account.

Some groups would develop their own sophisticated system for successfully adapting to the network rules. Others would learn the hard way, after painful and frustrating experiences. Some would be unable to adapt at all. And still others would experiment with ways of circumventing the system altogether.

Meanwhile, as the networks were developing political methods for controlling advocacy groups, industry lobbyists in Washington were working diligently to remove the legal powers these groups had recently won. In 1969, the same year that the first petitions to deny were filed by citizens' groups, the National Association of Broadcasters began pushing Congress for legislation to make it more difficult to challenge a license. Anticipating opposition from minority groups, the NAB hired a black consultant—whom they paid $2,500 a week—to rally support from black political organizations. But most minority media activist groups were clearly aware of the potential impact of the proposed legislation, and they began fighting it immediately. During hearings in Congress, members of Black Efforts for Soul in Television (BEST) picketed the Washington headquarters of the NAB as well as the network-owned and -operated stations in Boston, Chicago, Philadelphia, and San Francisco. "This bill represents backdoor racism," they told the press, "because it is a subtle, and therefore vicious attempt to limit the efforts of the black community to challenge the prevailing racist practices of the vast majority of TV stations."[36]

A coalition of media reform groups successfully prevented

the bill from passing. The following year, broadcast lobbyists persuaded the FCC to institute rules at the Commission that did basically the same thing the bill had been designed to do. Again media reform groups fought back, and the courts struck down the rules.[37]

But the struggle was not over. Broadcasters continued to press for legislative or regulatory relief from the threat of citizen groups, waging a public campaign against the media reform movement. NAB chief Vincent Wasilewski warned a group of broadcasters in 1974 that "pressure groups using the government process to manipulate programming to meet their own selfish needs pose as big a threat as government dictated programming." Added another broadcaster: "Peace in our time may have come for most Americans; it has not come for broadcasters. . . . We must fortify ourselves with sufficient ammunition and extensive legal armament for full-scale warfare."[38]

Invisibility and Influence

That Certain Summer was a breakthrough for prime-time television. Written and produced by the team of Richard Levinson and William Link, the ABC TV movie featured Hal Holbrook as Doug Salter, a divorced father whose son comes to stay with him for the summer. The boy is shocked to learn that his father is homosexual and lives with a lover (played by Martin Sheen). For two hours, the characters in this penetrating drama struggle with the issues of honesty, personal choice, and acceptance. Though the child is ultimately unable to accept his father's lifestyle, the movie is a sympathetic portrayal of what it means to be gay. Never before had this touchy subject received such serious and sensitive treatment on television. In fact, prior to the airing of this film on November 1, 1972, the topic of homosexuality had hardly been dealt with at all in prime time.[1]

Like the *Maude* series (which debuted that same autumn), *That Certain Summer* reflected network television's growing interest in controversial issues. But while *Maude* was criticized for presenting abortion in a comedic context, *That Certain Summer* was critically acclaimed for its serious and compassionate treatment of a delicate subject. It also did very well in the ratings. Its success suggested that provocative social and political issues might be the ideal ingredient for network television's newest genre—the made-for-TV movie. As one-shot broadcasts, television movies needed story lines that could draw audiences. *That Certain Summer* fit the bill very well, generating considerable press coverage in the weeks prior to its airing.[2]

But the film was important for another reason as well. By

ABC's 1972 television movie *That Certain Summer* was the first full-scale treatment of homosexuality in prime time. *(Courtesy of William Link and Universal Studios)*

bringing homosexuality out of its television closet, *That Certain Summer* helped make prime-time TV a target for influence by gay activists. For it wasn't until after this first serious treatment of homosexuality that gays began approaching the networks about their portrayal.

The gestation period for *That Certain Summer* coincided with a crucial time of renewed gay activism in America. The Stonewall Riots—which erupted in 1969 when police raided a gay bar in Greenwich Village—had launched the gay liberation movement. Shortly afterwards, gay activist groups began forming all over the country. By the time the film aired three years later, political activism among gays and lesbians had risen dramatically. The movie got mixed reviews from the gay community. Some praised it; others complained that it didn't go far enough. As one viewer wrote in a letter to the *New York Times:* "Why did the two male lovers never touch or kiss? Why did Doug Salter cry at the end of the film—his tears were a repudiation of the life he had chosen for himself. Why did his son reject rather than accept him?" Levinson and Link attribute this negative criticism to the rapid rise in militant activism among gays:

> The gay sensibility had altered during the time span between the conception of our film and its airing. When *That Certain Summer* was shown on television the militants had arrived at a point where they did not wish to be reminded that there were still many among their number who were troubled, unsure, and not quite ready to face society with a strong sense of self identity. The character of Doug Salter was a homosexual in transit, and therefore not as liberated as the militants may have wished. They wanted propaganda, not drama.[3]

It should not be surprising that the airing of a film like *That Certain Summer* would generate considerable expectations in the gay community. Like other advocacy groups, gay activists were beginning to see prime-time television as critical symbolic territory in their struggle to gain acceptance in the wider society. The film's success demonstrated television's power to bring an issue to national attention. It also showed that prime time was ready to deal with the difficult subject of homosexuality. Entertainment TV could become an essential political tool for advancing the cause of gay civil rights, just as it was for black, Hispanics, Asians, and women.

Gay activists shared many of the same goals and objectives of other media advocacy groups. They wanted wider representation in entertainment programs, as well as influence over the way they were portrayed. Like other groups, they sought: a way to get their views across to the right people in prime-time television, a knowledge of effective pressure points, and some form of leverage. But gays faced unique problems that set them apart from other advocacy groups. They were not afforded the kind of legal assistance that minorities and women got from the FCC. Nor did their cause have widespread support among the general population. Many people still viewed homosexuality as sinful, deviant behavior rather than a legitimate life-style. Thus, pushing for positive portrayals was much more difficult for gays than for women or minorities. Like abortion, homosexuality was the kind of explosive issue that could require television to walk a tightrope between opposing groups. Despite these obstacles, gays had one important advantage over other groups. They referred to it as their "agents in place."

According to gay activists, there were a substantial number of gay people working in the television industry who were not open about their life-style. Some held high-level positions. While unable to promote the gay cause on the inside, they could be very helpful to advocates on the outside, especially by leaking information. These "agents in place" became one of the linchpins of gay media strategy.[4]

Like the network system for managing advocacy groups, the strategy used by gay activists to influence prime-time programming evolved over time. Through trial and error, gay activists learned which tactics were most effective in dealing with the networks. They learned how to adapt their tactics to fit network strategies. Sometimes gay activists worked within this system, sometimes around it. While some of their efforts were modeled after what other groups were doing, gays ultimately worked out a method of influence that reflected their own needs, strengths, and goals. In time, the gay activsts gained a reputation within the industry as the most sophisticated and successful advocacy group operating in network television.[5]

Shortly after the groundbreaking broadcast of *That Certain Summer,* several gay activist groups began to approach the networks to discuss the portrayal of homosexuality. One was the New York-based Gay Activist Alliance (GAA). GAA's media director, Ron Gold, had been a reporter for *Variety,* so he had some familiarity with the decision-making structure at the networks. In January 1973, he wrote to all three network standards and practices departments, requesting meetings. Before a meeting had been scheduled with ABC, GAA members were smuggled a script by one of their agents in place. It was for an upcoming episode of *Marcus Welby, M.D.,* entitled "The Other Martin Loring," and it concerned a married man who asked Dr. Welby to help him with his homosexual tendencies. Welby assured the man that as long as he suppressed his homosexual desires, he would not fail as a husband and father.[6]

As Gold remembers, GAA leaders "blew a cork" when they read the script. They were particularly offended that the show was coming from the same network that had played an important role in promoting positive, sympathetic portrayals of gays. Instead of waiting for an appointment with ABC executives, the activists—with the help of another network insider—"took over" the network executive offices. Recalls Gold: "We knew somebody who worked there who gave us a kind of plan of the place and we did a little scouting in advance and we managed to sneak into the offices. The confrontation at ABC headquarters was hostile and explosive. This unexpected visit from twenty-five angry activists was hardly the manageable kind of meeting network executives preferred to have with advocacy groups. Executives offered to talk with two group members if the others would leave. But, as Gold recalls, "we said we would talk to them but we wanted everybody to stay there. . . . They wouldn't agree to that so everybody went out and people got arrested."[7]

This first angry encounter didn't keep the objectionable episode from airing a few days later, but it did have an impact on later decisions. As they were beginning to do with Justicia and other groups, ABC executives decided to invite gay activist comments on any new scripts dealing with homosexuality. Since gays had their own ways of getting scripts anyway, this approach was even more essential than with other groups. The following year, the producers of *Marcus Welby* submitted another script dealing with homosexuality. Entitled "The Outrage," it revolved around a male teacher who molested a teenage boy. This time, the network standards and practices department gave a copy of the script to the gay activists.[8]

Since the first meeting with ABC, Gay Activist Alliance had experienced an internal rift. Some of its members, including Ron Gold, separated from the group and formed their own organization. Called the National Gay Task Force (NGTF), it was set up as an umbrella organization for grassroots gay rights groups around the country. Both NGTF and the GAA then began to compete to become the "one voice" to represent gays to network television. ABC chose to submit the *Welby* script to the newer group.

By allowing the activists to see the script, network executives no doubt hoped for approval or suggestions for minor changes in the story that would make it less offensive to the gay community. But a story line that appeared negotiable to the executives proved to be unacceptable to the activists. The very premise of the episode—tying homosexuality to child molestation—was a prime example of what gay activists wanted to eliminate from the mass media. To them, the broadcast threatened not only to reinforce anti-gay sentiment but to create it as well. The situation was exacerbated by Ron Gold's explosive reaction. When he lost his temper with standards and practices executives, communications broke down between the network and the activists.[9]

The National Gay Task Force then turned the matter over to another activist, Loretta Lotman, one of the pioneer gay media activists in Boston. Since meetings with network executives had not been successful, Lotman and her colleagues decided to launch a nationwide grassroots campaign against the *Welby* show.

This first national campaign by the gay and lesbian community became a turning point in the evolving relationship between gay activists and the networks. It galvanized the national gay constituency, focusing their attention on a single issue. It publicized the gay rights cause, garnering support from sympathizers outside the gay community. It served as basic training, during which national and local gay activist leaders learned how to pressure the television industry. And, most important, like the *Maude* protest, it was a dramatic show of power. Though ultimately unsuccessful at keeping the program off the air, the campaign against *Marcus Welby* demonstrated that gay activists had the constituency and the know-how to apply considerable pressure on network television. Unlike those who protested over *Maude,* however, gay activists used this first incident to begin the development of a sophisticated system for influencing network television.

In their campaign against *Marcus Welby,* the gay activists planned to strike at the most vulnerable pressure points in the network industry. As the Catholic organizations had done the year before, the gay activists used their grassroots groups to apply pressure on local ABC affiliates. But gay groups had a decided advantage. They had advance knowledge of the upcoming episode, and therefore more time to organize their pressure efforts. In some cities, the gay activists knew more about the controversial network program than the station executives did.

Like other media activists, Loretta Lotman already had established ties with the management of her local Boston ABC affiliate, WCVB. She telephoned the station a few weeks before the scheduled broadcast of the *Welby* episode. When asked if he knew what the network planned to "foist" on him, WCVB's program director was caught by surprise. Lotman urged him to find out about it right away, warning that, if something were not done about the program, the station would be "hit with a protest the likes of which you've never seen before." Lotman then contacted gay and lesbian groups throughout the country, sending them copies of the script, along with detailed instructions on how to pressure affiliates. Knowing that local activist leaders would encounter skilled community liaison people at the stations, Lotman shared some of the lessons she had learned in dealing with station personnel. "Keep your temper," she advised. "We can yell at each other. You have got to get a meeting there. They're going to try and give you a tour of the station so you'll be so in awe of the 'Great God Media' that you'll be incapacitated in the meeting. Don't buy it. Go get a tour beforehand if you have to get yourself immuned, but go in there and tell them they're in a very bad position."[10]

Pressure was also directed at advertisers. Through their agents in place, gay activists were able to find out which companies had bought advertising spots in the upcoming episode. As Lotman remembers, "Certain compacts were made with people who were broadcast professionals at the ABC network, including some people in some very well placed positions. I can go no further than that." The National Gay Task Force then instructed its members to write protest letters to these advertisers. NGTF also used the gay press, which was already regularly carrying stories about the Welby campaign, to publish the names of the sponsors.[11]

Activist leaders used the mainstream press to publicize their campaign and to generate support from people outside the gay community. They also made specific appeals to professional organizations. Only a year earlier, gay activists had participated in a protracted struggle within the prestigious American Psychiatric Association to get homosexuality removed from the organization's list of mental illnesses. Having succeeded in that effort, NGTF now convinced the APA to issue a public statement condemning the *Marcus Welby* episode. Gay leaders also persuaded the National Education Association to put out a press release objecting to the show's negative portrayal of the teaching profession, as well as its "misconceived, stereotypical portrayal of a homosexual [which] may deepen public misunderstanding rather than enlighten or educate in any way."[12]

As a counter-strategy to this well-publicized campaign against the *Welby* show, ABC issued its own statements defending the episode, asserting that a psychiatrist had been consulted on the preparation of the script. At the same time, the network and the producers made changes in the program in an attempt to minimize its offensive elements. During this last-minute, patch-up effort, several scenes were removed from the controversial episode and some material was reshot; overt references to homosexuality were deleted; and the term "pedophile" was introduced as the new label for the sex offender. Although the network had identified the episode with homosexuality a few months earlier—and hence had initiated contact with the gay activists—officials later explained to the press that the program was not about homosexuality, but about child molestation and the "extreme emotional problems physical assault can cause its victims."[13]

These stopgap measures proved to be too little and too late to reverse the effects of the protest on advertisers and affiliates. Seven of the companies that had bought time in the one-hour drama withdrew their ads, leaving only one minute of air time sold before the broadcast. And at least five affiliates refused to air the program. The pressure campaign had drawn attention to the controversial nature of the episode, which many advertisers preferred to avoid. It was one thing to buy into shows with provocative content, quite another to subject one's company to the damaging association with such a contentious and unsavory plot line. The ABC stations in Lafayette, Louisiana, and Springfield, Massachusetts, both announced their refusal to air the *Welby* episode because the subject matter was "unfit for prime-time viewing."[14]

But in the larger cities, especially those with sizable gay communities, local station executives were more directly influenced by the gay rights activists. WCVB's general manager in Boston told the press that he had been impressed by the "quality of letters" protesting the show. The episode was being rejected because of fear that it would reinforce the notion that homosexuals are commonly child molesters. "That isn't true," he carefully explained to reporters, "but that's what would come through to an audience." In Philadelphia, the management at WPVI-TV issued the following statement, explaining their reasons for refusing to air the episode:

> It appears to us from the outset of this program that integral to the author's original premise is a false stereotype of homosexuals as persons who pursue and sexually assault young boys.

> While it is also clear that the producers have earnestly attempted to alter the effects of this unfortunate premise, it is equally clear that they have not succeeded. We think, also, that the presentation contains substantial improbabilities and contradictions that impinge upon the credibility of the situation portrayed while actively reinforcing the negative stereotype.[15]

Since the New York station was owned by the ABC network, its management could not as easily reject the program. However, to cover itself under the Fairness Doctrine, WABC agreed to broadcast a pro-gay documentary, entitled *Homosexuality: The Open Secret,* in which gays were given the opportunity to "speak for themselves with a minimum of interference from the narrator." The documentary's strong conclusion declared that "all of our institutions must sooner or later adjust to the idea that homosexuality may be nothing more than a normal variant in the total spectrum of sexual behavior."[16]

ABC's decision to air the *Welby* episode came as a surprise to the leaders of the protest campaign. Such vociferous opposition from the national gay community seemed to them enough to persuade the network against the broadcast. National Gay Task Force leaders charged that their efforts had been sabotaged by several members of the Gay Activist Alliance, who secretly and unilaterally offered ABC their approval of the show on behalf of the gay community. But ABC executives obviously had other factors to consider as well. Even with the pullout of sponsors and the loss of some affiliates, not to air a program that had generated so much press attention would appear to be capitulation. This could set a precedent that would alarm advertisers and affiliates alike. Weathering the strong reaction of gays after the broadcast might have seemed an easier course for the network in the long run.

Gay activist reaction to the October 8 airing of "The Outrage" was predictably vehement. A coalition of groups picketed the ABC station in Washington, D.C., some of them dressed up as "Dr. Marcus Quackby." The National Organization for Women participated in the protest, complaining that several "male chauvinist stereotypes" appeared in the program along with the homosexual stereotypes. National Gay Task Force representatives wrote angry letters to ABC-TV president Elton Rule, chastising the network and demanding that the offensive program not be aired again. This time ABC agreed, and the episode was quietly withdrawn from the reruns.[17]

Though this campaign was not entirely successful, it did have an impact. Because of advertiser drop-out, the episode lost money for the network. But, more important, the protest campaign had shown the TV industry that the gay activists were capable of causing major disruptions when they objected to programming. This could have worked to the disadvantage of gays. ABC might have concluded that it would be easier to leave the subject of homosexuality out of prime time altogether than to face the wrath of unhappy activists. Leaders of the gay community knew this. So, to ensure continued incorporation of gay issues into entertainment programming, they began developing a coordinated strategy.

Influencing a large, complex institution like network television required not only sophisticated political skills but an understanding of the structure and operation of the industry. As other successful advocacy groups would do, gay activists had to educate themselves in the workings of network television.

One of the most important lessons from the *Welby* confrontation was that divisions among groups could too easily weaken their position. In the years following this first protest, the National Gay Task Force became the official group to deal with the networks on behalf of the gay community. Though the movement itself was not monolithic, and from time to time there were serious disagreements among the various divisions within it, gays and lesbians, in their media strategy, consciously sought to behave as a single political entity. An NGTF pamphlet for local groups explained the rationale for this unified image:

> Before meeting with you, the media will want to be sure that they aren't going to be bombarded with similar requests from competing gay organizations. They want to feel that they will meet with representatives of the gay community, not just with a few individuals who only represent themselves. So it's a good idea to establish your credentials in a letter. If you're the only group in the area, or the only one that's active politically, that's good enough. But, if you're not—and particularly if there are

> not women or men in your group—it would be advisable to approach other gay groups in your community and send a letter signed by leaders of two or more of these organizations.[18]

As important as a unified image was the need for the activist group to be located near the centers of decision making. Shortly after the Welby protest, Loretta Lotman moved to New York and became the full-time media director of NGTF, establishing regular ties with the executives in the standards and practices departments at all three networks.

But the gay activists soon learned that being in New York still left them at a disadvantage. Not only was most TV production done in Los Angeles—as East Coast standards and practices executives routinely explained to complaining advocacy groups—but the networks themselves divided their decision making between the two cities. While overall policy decisions were made in New York, the day-to-day programming decisions took place on the West Coast, where both standards and practices and programming departments had large staffs. When network executives invoked this 3000-mile separation as a way to shift responsibility to more distant areas of the industry, the gay activists quickly came up with their own counterstrategy.

In order to make the networks accountable on both ends of the continent, the National Gay Task Force encouraged its California allies to open up a "West Coast branch" in Los Angeles, which they called the Gay Media Task Force. Though the two organizations were not officially tied together, they operated as a team when dealing with the networks. The Gay Media Task Force was really a one-man operation, run by psychologist Dr. Newton Deiter. Working out of an office in his home, Deiter spent some of his time counseling patients and some of it consulting with the networks on gay issues.

The bi-coastal set-up gave the gay activists an edge over other groups who were based in only one city. It became virtually impossible for the networks to escape the scrutiny of gay activists. In addition, grassroots activist groups functioned as "affiliates" in the gay lobby. These local groups could be called upon to pressure local stations if necessary, as they had done in the

Welby protest. This overall structure paralleled the geographic and decision-making structure of the network television.

Rather than appearing to be a demanding pressure group, the advocacy group leaders presented themselves to the networks as a "resource" for information about homosexuality. Referring to itself as an "educational lobby," the National Gay Task Force regularly provided media decision makers with statistics and research to dispel commonly held myths about homosexuality. NGTF representatives repeatedly asserted to the networks and to the public that the group had no intention of censoring program content. What they were doing, they insisted, was simply helping the networks in their own self-censorship process.

But the advocacy group did not restrict itself to occasional network-initiated technical consultation. NGTF had an agenda for specific changes in network programming. This was made very clear to the network standards and practices departments on a regular basis. Arguing for "minority group status," gay activists demanded: increased visibility, elimination of stereotypes, continuing gay and lesbian characters, and gay couples. Gays also insisted on a "moratorium on negative portrayals."[19]

In contrast to some of the advocacy groups that had approached the networks with a strong stance and then vanished when funds or organizational momentum ran out, NGTF set itself up as the most prominent national gay organization in the country. With funding from foundations and individuals, NGTF employed a full-time media director whose job it was to initiate and maintain regular contact with media organizations. Gays thus became an ongoing political presence in network television.

This presence was enhanced and supported by the infrastructure of gays working in the industry. Insiders continued their surveillance of the television industry, assuring that the networks carried out the agreements made with the activists, and keeping a watchful eye for any offensive content that might have slipped by the scrutiny of the standards and practices departments. Sometimes the gay activists knew about problems before the network censors had even seen them. In one case, an agent in place, working on the set of an NBC situation comedy, telephoned Newton Deiter about a problem he'd spotted during rehearsals. One of the actors was behaving in a noticeably "limpwristed" fashion, he reported. Within minutes, Deiter was on the phone to the manager of the Los Angeles standards and practice office, who was surprised to learn of the incident. Such instances of surveillance encouraged the networks to consult more consistently with the gay lobbyists at the script stage. If such consultation did not occur, gays were often able to find out about the scripts anyway. As Deiter noted, "They finally learned over there that we find out."[20]

The monitoring of industry operations was paralleled by a continual monitoring of programming content. Other advocacy groups used different kinds of analyses of program content. The National Organization for Women commissioned various studies that identified the number of women and the kinds of portrayals in prime-time television. These quantitative studies were often used to educate decision makers at the network on the overall representation of a group or issue. Gays engaged in a more grassroots form of monitoring. Since gay activists had an inside track on most gay portrayals before they were broadcast, gays around the country could be alerted by newsletter of upcoming programs to watch. They were encouraged to send their reactions back to NGTF. They were also advised to report any other gay portrayals they came across on television. These grassroots reports in turn would be presented as "feedback" to the network by NGTF's media director. Sometimes the feedback would be part of a periodic meeting which NGTF would call with network executives; at other times, the comments would be part of a letter from NGTF's media director. These reports were always concrete, carefully presented, and balanced. Portrayals which gay activists approved were lauded, and specific reasons were given for what was good about them. Objections to negative presentations were also explained in detail. In this way, the gay activists were sensitizing network decision makers to the nuances of behavior of which gays approved or disapproved.[21]

In a 1978 meeting with CBS, NGTF leaders summarized the accomplishments of the network in portraying homosexuality in its programming by presenting to executives a list of portrayals, stereotypes, and images that the group considered "Good News" and "Bad News." "Stereotypes" to which gays objected included:

> murderers, child molesters (male prostitution); mental disturbance (pathology); weak or absent father, domineering mother, unhappy childhood; bad experience with opposite sex—promiscuity—no lasting relationships, unfulfilled, miserable empty lives. Gay men = swishy, limp-wristed, female role, want to be women, transvestites, transexuals, Instant hilarity. Lesbians = masculine, want to be men. No comic value.

"Good images" included:

> person doing a good job—gay cop, business executive, sportsperson, secretary, psychiatrist—*mainstream*. Person who stands up for himself/herself, people of courage; heroes sensitive, compassionate, ethical, personable. Loving and affectionate gay couples. Gayness just *incidental*. More *lesbian* portrayals.[22]

While presenting themselves as cooperative lobbyists, the gay activists also made it clear that the possibility of a protest was never out of the question. The Welby protest was periodically invoked—especially to ABC—as a threat if the network would not cooperate with the activists. When ABC failed to consult with the advocacy group on an upcoming TV movie, NGTF wrote a letter to executives suggesting that grassroots groups were poised for another battle with the network:

> The way your network representatives are handling the "Jenny Storm Homicide" is leading directly to another Welby-ish confrontation. Such a situation is not to be desired. We don't even know if the material is offensive or not. However, you are giving us the impression that ABC has something to hide from us. Going from past experiences, we're not going to wait for it to air to find out whether you've struck another blow against gay civil rights. . . . [W]e are prepared to move on 72 hours' notice. We will *only* if there is no other way.[23]

In this case, the "zap"—as the gay activists called their protests—was avoided. But there were occasions when gay activists did battle with the networks. These confrontations occurred infrequently—there were seven of them between 1974 and 1977—and were gradually reduced in intensity over the years. They usually resulted from some breakdown in the smoothly running relationship between the advocates and the networks.

One of the more memorable "zaps" occurred over a 1974 episode of *Police Woman* on NBC. Entitled "Flowers of Evil," the program featured three lesbians who ran a rest home and systematically murdered its occupants. National Gay Task Force leaders met with NBC executives in New York following the broadcast and demanded that it not be rerun. A few weeks later, then Vice President of Broadcast Standards, Herminio Traviesas, was attending a network meeting in Jamaica when he received a frantic phone call from his New York office. As the corporate executive recalled: "I was told that a group of lesbians had invaded my office and would not leave unless we guaranteed them 'Flowers of Evil' would not be rerun." Members of the New York-based Lesbian Feminist Liberation Organization—alerted by members of NGTF—had used an inside contact in order to sneak into the well-guarded NBC headquarters in Manhattan. Some of them had brought their babies with them. Traviesas ordered his staff to feed the women and let them stay for twenty-four hours until he could catch a plane to New York. Although no agreement was reached at the meeting, the National Gay Task Force received a call from NBC a short while later, assuring them that the episode would not be rerun, and it was not.[24]

Disruptive incidents like this one provided additional incentives for networks to consult with gay representatives during the development stage of TV programs. For that process, the Los Angeles-based Gay Media Task Force became the central clearinghouse. Newton Deiter was designated as the primary technical consultant to the networks on any gay-related program material. He also functioned as the hub of a network of more specialized gay consultants, who could be called upon for input. If a script needed the help of a gay or lesbian teacher, lawyer, or psychologist, Deiter would recommend the appropriate consultant. Deiter himself not only consulted regularly on scripts with gay characters or issues in them but also sought to encourage incorporation of gay characters into programs where there otherwise may not have been any. These consultation services, for which fees were sometimes paid, ranged from a quick telephone call to get a reaction on a line of dialogue to actual participation in the scriptwriting process. The services of the Gay Media Task Force were generally well received by the production community. Producers found the technical assistance of Deiter and his colleagues helpful not only in providing information on homosexual issues, but also in getting otherwise politically volatile material on the air with a minimum of protest from the gay community.[25]

In his dual role as consultant and advocate, Deiter was in a curious position. As a spokesman for the national gay community, he was expected to take a strong position on portrayal of gay characters and gay issues in order to ensure that representation of gays on television was consistent with the objectives and policies of the gay activist leadership. At the same time, Deiter was working within an industry with its own imperatives and constraints, which were sometimes in opposition to the expectations of the activists. To function effectively in Hollywood, Deiter had to learn how to make compromises. He also quickly internalized the rules of the game for the production of prime-time entertainment programming. In many ways, Deiter's dealings with television producers paralleled those of the network standards and practices departments. Just as the editors in standards and practices departments carefully couched their requests for script changes in "helpful" terms, Deiter applied the same principles. Refraining from a heavy-handed, censorial approach, Deiter never told producers they could not do something. Rather, he explained, (using the very same phrase used by standards and practices executives): "We always try to come up with alternative methods to accomplish what they [the producers] want." While he always appeared conciliatory and cooperative with producers and networks, Deiter often used his connections with the National Gay Task Force as leverage to encourage compliance with his recommendations. As the official barometer of gay attitudes and responses, the consultant would periodically suggest that, though he may not personally object to this line or that portrayal, he could not guarantee that his constituents would be so agreeable, and he was obligated to inform the National Gay Task Force of whatever decision the network made. These warnings, with their veiled threats of possible "zaps" from the gay activist community, provided additional weight to the consultant's recommendations.[26]

Sometimes Deiter's compromises did not meet with the approval of gays at large, or with his colleagues at the National Gay Task Force. While the NGTF insisted on a moratorium on negative portrayals of gays, Deiter's policy was to allow a negative character to appear in a program, if there were a positive one to balance it. In one case, Deiter approved the incorporation of a limpwristed swishy gay purse snatcher and his lover in an episode of *Barney Miller.* The consultant saw the two characters as very funny parodies of familiar stereotypes. But many gays were offended, and they wrote angry letters to the National Gay Task Force complaining about the program. Deiter then went back to the show's producers, and persuaded them to do another episode which challenged prevailing myths about homosexuality.[27]

Deiter's intervention in specific programs was sometimes quite extensive. The consultant worked closely with the producer and writer of two 1979 TV movies about male prostitution: *Dawn: Story of a Teenage Runaway* and its sequel, *Alexander: The Other Side of Dawn.* In fact, according to Deiter, one of the characters in the drama was modeled after the consultant himself. On other programs, Deiter succeeded in reversing the plot line so that it would reflect a pro-gay point of view. In the script for a 1977 episode of *The Streets of San Francisco,* criminals tried to blackmail a closeted gay cop. When the cop went to his superiors with the truth, his partner refused to work with him. The end of the drama showed the gay policeman resigning from the force, while making an impassioned speech about intolerance. Deiter felt the script was sending a negative message. He pointed out to producers that the real police chief in San Francisco had urged cops to come out of the closet. In compliance with Deiter's suggestions, the script was rewritten. In the new version the partner conquered his homophobia, the gay policeman kept his job, and the two of them continued to work happily together.[28]

Gay activists were so successful at establishing themselves as a presence in the television industry that more and more gay characters began to appear on prime-time TV screens throughout the seventies. These portrayals clearly reflected the consistent input of the gay lobbyists. "All but gone are lisping gays and homosexual murderers and child molesters," noted one TV critic in 1982. "Virtually every series has done its gay show." But the pattern of gay representation in entertainment programs also reflected the compromises that had to be made in incorporating the controversial issue of homosexuality into a commercial, mass medium. Though the advocates used every possible means to push their agenda as far as it could go, the boundaries of network television shaped the portrayal of gays and gay issues.[29]

In 1976, which one critic labeled "the year of the gay" in television, these patterns were clearly in evidence. Gay characters appeared in at least seven situation comedies and in several television movies that year. All of these programs involved some consultation with the Gay Media Task Force. Most of the characters appeared one time only in the sitcoms and vanished the following week. Generally the focus of the plot was on the acceptance of gay characters by the regular heterosexual characters. Very few gay couples were shown, and they were not permitted to display physical affection.[30]

In an episode of *Sirota's Court*, the judge agrees to perform a wedding ceremony for two gay men. Though acknowledging that marriages between gays are illegal, he explains to the court that he is "testing the law." At the end of the ceremony, the judge pre-empts a possible kiss of the two newlyweds, and orders them to shake hands.

In the television movie *In the Glitter Palace*, the subject of lesbianism is examined in the context of a conventional crime drama. Because of extensive consultation with the Gay Media Task Force, the producers went to great lengths to make the portrayals of the two women positive. At the beginning of the film, the audience is purposely shown that the lesbian accused of murder is innocent. But the two women lovers are only allowed the very minimum of affection between each other. One of them spends much of her time in jail. The one embrace between the two women could easily be interpreted as the innocent hugging of two heterosexual girlfriends.[31]

In situation comedies that season, the comedic device of mistaken identity became a convenient way to tie in the subject of homosexuality. On one episode of *Alice*, a husky, athletic fellow is Alice's date for the evening. She can't figure out why he hasn't made a move on her. Finally, he informs her—quite proudly—that he is gay.

Most of the shows that featured gays that season appeared to be conscious efforts at public education. In virtually every one of them, the heterosexual characters learn to accept gay people and their life-styles. The dialogue in many of these shows includes some rather self-conscious sermonizing. Phyllis, of the popular show by the same name, finds herself on a date with a guy who turns out to be gay. Steve confesses to Phyllis that no one else knows he is gay. He wants to tell his parents, but is fearful of their reaction. "These are the 1970s," Phyllis assures him. "Being gay isn't something you have to hide anymore." Phyllis encourages him to tell his parents, but he chickens out and announces to them that he and Phyllis are engaged. It isn't until the "engagement" party that Steve gets up the nerve to tell his family the truth. When he finally does, they immediately accept him and he goes around the party with Phyllis in arm, announcing, "I want everyone to know that I am gay and I have this woman to thank for it."

On *CPO Sharkey*, the audience is introduced to the new terminology. Says one character: "Do you know what it means to be gay?" The second: "Do you mean like on New Year's Eve?" The first: "No, to be gay means that on New Year's Eve you'd rather be with Burt Reynolds than Debbie Reynolds."

Treatment of homosexuality in television shows sometimes reached the point of self-parody. Richard Levine described one show which included:

> a dispute between network censors and a TV producer about whether certain jokes in his show would offend gays. At the height of the argument, the producer brings out the head of the "Gay Task Force" from the next room. Mincing and flouncing in the most stereotypical manner, the gay leader walks in, limpwristedly shakes hands all around, and says, in a lisping,

> high-pitched voice, "We're not really concerned with occasional derogatory emphasis." Then his voice drops two octaves and his effeminate mannerisms suddenly cease. "I'll tell you what we do find offensive," he continues. "We are deeply offended by the fact that supposedly sophisticated men like you could so readily accept the fact that a gay person would come in here talking like Sylvester the Cat."[32]

By the mid-seventies, gay activists had become so institutionalized in network television that they rarely needed to use protests. Relations were smooth between gays and the networks. More and more the media work of gay activists shifted from the East Coast, where they had first gotten access to network decision makers, to the West Coast, where the Gay Media Task Force had become part of the fabric of the production community. Though representations of gays in prime time would always remain circumscribed by the constraints of commercial television, gays would continue to make progress. But other forces were beginning to mobilize which would weaken the influential position gays had fought so hard to establish. As gays and lesbians became more visible—both in programming and as an activist group—they would draw more attention from political conservatives who would exert counter-pressures on the network television industry. The 1977 controversy over *Soap* signaled some of these new developments. Just as CBS had found itself in the middle of a collision between liberal and conservative forces over *Maude,* so ABC placed itself in a similar crossfire with *Soap.* Though the gay activists were firmly entrenched in the TV industry, they found the task of negotiating their portrayal in this case more difficult than usual.

Soap was in hot water before it ever got on the air. One of programming chief Fred Silverman's pet projects, *Soap* was a prime-time spoof of daytime soap operas. Touted as the most important breakthrough since *All in the Family,* the new series vowed to leave virtually no controversial subject untouched. A memo from the standards and practices department described the show as "a further innovation in the comedic/dramatic form presenting a larger-than-life frank treatment of a variety of controversial adult themes such as: premarital sex, adultery, impotence, homosexuality, transvestism, transsexualism, religion, politics, ethnic stereotyping (and other aspects of race relations), etc." When it was screened to station executives during the yearly affiliates meeting, many were "appalled" at the new program and one of them referred to it as "one long dirty joke." When *Newsweek* published a story about the affiliates' reaction to the new series, *Soap* became the subject of a public controversy within days.[33]

Most of the protest over *Soap* came from religious organizations. Four powerful church groups—the National Council of Churches, the United States Catholic Conference, the United Church of Christ, and the United Methodist Church—representing 138,000 member churches, were particularly active in organizing protests against the show. In a little over four months before the show premiered, ABC received more than 22,000 letters, mostly negative. In the midst of the uproar, a number of advertisers withdrew their ads before the show was aired, and at least twelve affiliates refused to run the first two episodes.[34]

While the churches protested many elements of the new series, gays were concerned about one thing: Jodie, the show's continuing gay character, the first in prime time. Newton Deiter was given a private screening of the first two episodes of *Soap* before they were shown to affiliates. In the episodes—which Deiter was told were an "in-house test"—Jodie was portrayed as a swishy stereotypical homosexual who liked to try on his mother's clothes and who desired a sex-change operation. Though Deiter wasn't particularly happy with the character, he didn't make objections. Since the show presented a unique opportunity to develop a continuing character, the consultant offered suggestions on how that might be done. In a letter to ABC's West Coast broadcast standards chief, Deiter wrote: "I understand what the producers are trying to do in that no one will be sacred in this presentation." He then went on with several pages of detailed recommendations on how the character could be changed so that he would not be "grossly offensive to the gay community." "In the next episode to be filmed," Deiter proposed that Jodie "discard the idea of being a transsexual and, instead, become an upfront and somewhat militant Gay Liberationist. The advantages are several," Deiter explained to the network executive. "The characterizations and slurs that are so much a part of the show could then remain, except 'Jodie' would be answering back in kind. That could possibly result in his emerging as an even stronger character." The consultant ended his letter with a mildly stated—but nonetheless clear—threat: "We stand ready to be fully cooperative in keeping this character funny, outrageous and real," he wrote. "[However,] should the character remain as shown in episodes 1 and 2, there will undoubtedly be substantial backlash and reaction from the gay community nationally. At this point, the matter need not go in that direction, since letting our constituency know of the character's future development, we should be able to assist aborting any protest."[35]

If the show had not already gained so much notoriety, the network might have willingly complied with Deiter's requests. But a great deal of attention was already focused on the new series, not just from the gays but from religious organizations that were outraged at what they had heard about the *Soap*'s immoral content. To assuage the nervousness of affiliates and advertisers, Fred Silverman told the press that developing plot and character lines would not be "immoral" and that Jodie was going to "meet a girl and will find there are other values worth considering."[36]

Gays around the country were enraged at this statement, which seemed to indicate a complete reversal of progress for them in prime time. Deiter continued to negotiate with standards and practices executives over the development of the gay character, urging the National Gay Task Force to "hold off on a protest" until after the series had begun. Deiter assured his colleagues that, despite what Silverman was saying publicly, network executives were encouraging the producers to modify the character in a way that would meet with the gays' approval.[37]

But, when leaders of the NGTF held their own private screening of the yet-to-be broadcast first two episodes (which they obtained through a gay sympathizer in a New York ad agency), the activists were not as understanding as their Los Angeles counterpart had been. NGTF's media director, Ginny Vida—who had taken over the position from Loretta Lotman—

telephoned the network's New York offices, and "threatened another *Marcus Welby*" unless substantial portions of the show were deleted. Unsatisfied by the response from the network, NGTF took out an ad in *Variety* to publicly protest the series. The one-page advertisement was headlined: "Why One of the Largest Ad Agencies in the World Will Not Let Its Clients Sponsor *Soap*." The ad agency was never identified but was said to have refused to let its clients sponsor the show because of the program's stereotypical and negative treatment of gays. The copy read in part:

> We of the National Gay Task Force are particularly angered by a gay character on "Soap" who is portrayed as a limp-wristed, simpering boy who wears his mother's clothes, wants a sex-change operation and allows everyone to insult him without a word of response. You know, a "faggot." We are angry that a national network could be so insensitive to 20 million people in their struggle for their rights. We are angry that a gay "Stepin Fetchit" is being trotted out for a cheap shot at easy humor. And, we are sickened that ABC finds the notion "hilarious." What we want is for the scenes involving the gay character to be reshot, as several scenes already were to appease others and their morality. We want reassurance that Jodie, the gay character, isn't going to go "straight"—as Fred Silverman implied when he told ABC affiliates that Jodie was going to "meet a girl and find there are other values worth considering."[38]

The ad appealed to members of the television and advertising industry to encourage their ad agencies to boycott the program. It also suggested that individuals write to Silverman and do what they could to get local affiliates to drop the program. NGTF also issued a "Gay Media Alert" to member gay groups around the country, urging them to put pressure on affiliates, go to the press, and write to advertisers and the network. Vida wrote a very strongly worded letter to Fred Silverman, in which the media director outlined NGTF's objections to the program. In reference to the possibility that Jodie was going to fall for a woman, Vida concluded the letter with this threat: "If this means that, in an effort to pacify the homophobes, *Soap* is going to engage in the fiction that all gay people need to change their ways is to meet 'the right woman' or 'the right man,' you are going to have the gay community down on your corporate necks in a way you've never experienced. (This is not a threat; it's a fact.)"[39]

Despite the intensity of these pre-broadcast protests from the gay community, the first episode of *Soap* aired without changes on September 13, 1977. Ad cancellations and affiliate drop-out did occur, but the protests from gays may have had little to do with it, since the program was under fire from so many other quarters.

In an exchange of letters with the network following the broadcast, NGTF continued to make threats. East Coast vice president for broadcast standards Richard Gitter responded to NGTF's letter to Silverman with one of his own, in which he defended the portrayal of Jodie, characterizing him as "neither a transvestite nor a transsexual" but "a strong, positive character, comfortable in his sexual preference . . . [whose] mannerisms are neither stereotyped nor offensive. He invoked the serial nature of the program as providing opportunity for growth and change, and suggested that after NGTF had "viewed several episodes of our series, we will be happy to entertain your reactions and suggestions."

Gitter also made a statement that indicated some new directions for network policy. "ABC is receiving conflicting messages regarding homosexuality," the executive explained. "On the one hand, certain church groups criticize us for too positive a portrayal of homosexuality in 'Soap,' while you argue that our portrayal is negative and stereotypical, and likely to constitute a set back in the Gay Rights movement. The weighing of widely divergent points of view is illustrative of the problem we face in dealing with special interest groups. We take pains to ensure that our programming does not espouse a point of view, particularly as regards infidelity, promiscuity and homosexuality, which we recognize are extremely sensitive issues."

Though it had threatened to do so, NGTF did not launch a major campaign after *Soap*'s debut. The group may have succeeded with its earlier threats in letting the network know that gays would be closely watching future developments in the series. And the activists later claimed victory for their efforts, attributing positive changes in Jodie's character to their pressure.

Ginny Vida later recalled, "The character actually improved enormously. It turned out to be one of the more sympathetic and really sort of sane voices in the series."[40]

But the experience over *Soap* had troubled the activists. They were particularly alarmed at the network's response to pressure from conservative groups. If this kind of pressure intensified, gays might face new challenges in their efforts to maintain control over their prime-time image.

CHAPTER SIX

He Who Pays the Piper

Ira Davidson was shocked to hear the news. A veteran TV writer with a long, successful career in the business, he had never experienced anything like this before. A TV movie he had written was under attack for causing the death of a twelve-year-old boy. *Web of Fire* was about a deranged architect who set fire to his own building. It seems a young boy in Seattle had watched the movie and, in "copycat" fashion, had gone to his school, set it on fire, and died of smoke inhalation. Davidson, a responsible writer with a family of his own, was so disturbed to hear of the boy's death that he decided to fly to Seattle and find out for himself what really happened.

So goes the plot of *The Storyteller,* an NBC TV movie. Written and produced by Richard Levinson and William Link—the team responsible for *That Certain Summer—The Storyteller* was part docudrama, part mystery. For two hours, as the protagonist investigates the boy's death, the issue of TV violence is debated. Fictional social scientists cite research findings about the harmful effects of televised violence. Fictional TV executives defend the medium. And fictional "man-in-the-street" characters spout off their assorted opinions to the camera. The debate is purposely never resolved. In the final shot, a myriad of voices is heard on the sound track in a virtual cacophony, symbolizing the confusion and lack of resolution surrounding the violence issue. But in counterpoint to this ambiguity, the story itself absolves TV of responsibility. After his long search, Davidson finds out—much to his relief—that the boy was a troubled child. The parents had failed to seek psychiatric help for

Martin Balsam as a television screenwriter in a scene from *The Story-teller*. *(Courtesy of William Link and Universal Studios)*

him, though strongly advised to do so. "From the day we brought him home from the hospital," the boy's father admits, "he never seemed very happy."[1]

This was an unusual piece of entertainment TV. Seldom did prime-time programs deal seriously with television itself. Indeed, TV characters were hardly ever shown watching TV. It was even more rare for a show to tackle an issue as controversial as the effects of television violence.

Loosely based on an earlier TV movie, *The Storyteller* had its genesis during an unusual period for network television. When the movie was broadcast in 1977, the networks were under pressure from the most intense and broad-based campaign in their history. Organized forces had found a new and effective means of forcing the television industry to reduce violence. This well-publicized campaign used a sophisticated strategy to circumvent the network system for managing advocacy groups.

By the early 1970s, TV violence had already been the subject of controversy for more than twenty years. Social scientists and TV critics had researched and debated it periodically for years. Numerous congressional hearings had been held since the 1950s. Despite repeated promises of industry self-regulation, however, violence had remained a disturbing presence in prime time. The 1972 Surgeon General's Report and the hearings that followed it gave TV violence further prominence as a public issue. The report itself was tentative in its conclusions. The networks had managed to get some of their own people appointed to the research staff, effectively weakening the study's outcome. However, the testimony of the Surgeon General himself was unequivocal. "It is clear to me," he reported to Congress, "that the causal relationship between televised violence and antisocial behavior is sufficient to warrant appropriate and immediate action." Thus, as Geoffrey Cowan points out, "by the fall of 1974 it was widely agreed that the Surgeon General had concluded that television violence may be dangerous to your health."[2]

If the public needed more to rally around than a government report, they got it with the airing of an NBC TV movie in September 1974. *Born Innocent* starred Linda Blair as a young girl who is sent to a reform school for runaways. One particular scene—which just barely made it past the censors—was strongly attacked. As Cowan described it:

> Linda Blair—happy and serene in momentary isolation from the rest of the girls—is jarringly assaulted by a group of young inmates who proceed to destroy her innocence as they pull her out of the shower, throw her to the ground, force her legs apart and 'rape' her with the long wooden handle of a Jonny mop. The scene is filmed in graphic detail. As the wooden handle goes in and out of the girl, Linda Blair's anguish and screams are fully recorded in a performance that matches her most dramatic moments in *The Exorcist*.[3]

The broadcast elicited a huge public outcry. NBC received more than 3,000 calls and letters, running 20 to 1 against the movie. A few days after the airing of *Born Innocent,* a group of children in San Francisco committed a similar crime against a nine-year-old girl. The parents filed suit against NBC for inciting the crime. While the Surgeon General's report stimulated public discussion of television violence, *Born Innocent* and the violent crime that followed its broadcast dramatized vividly the urgency of the issue.[4]

Congress put pressure on the FCC, which in turn put pressure on the three networks. Industry leaders took immediate steps to institute self-regulatory measures, turning again to the Television Code for protection. CBS initiated the concept of a "family viewing hour" during prime time. That policy was then adopted by the National Association of Broadcasters, who made it part of the Code in February 1975. The new provisions stipulated that, between the hours of 7:00 p.m. and 9:00 p.m., "entertainment programming inappropriate for viewing by a general family audience should not be broadcast." The rule also required advisories to precede certain types of programming in later hours, warning viewers of possible "objectionable content."

But the Family Hour created more problems than it solved. The Hollywood creative community immediately opposed it because it imposed stricter censorship on programs, particularly the comedies with socially relevant content such as *All in the Family, M*A*S*H*,* and *Barney Miller*. Various members of the production community filed suit against the FCC for pressuring the networks, and they eventually won their case.

The new policy also failed to mollify the critics of television violence. Months after the NAB adoption of the Family Hour, the FCC was still being flooded with angry letters. One viewer urged the commission to take more drastic action, lamenting: "We're being exposed to the worst kind of cruelty and sadism. There's no place to hide."[5]

During government hearings on television violence, experts encouraged citizens to begin pressuring television advertisers. Testifying before the House Communications Subcommittee, psychologist Alberta Siegal recommended that "consumers convey their disapproval of violence vendors in two ways. We may refuse to purchase their products. And we may refuse to buy stock in their firms." By mid-1975, it appeared that the public was taking this advice seriously. As the trade publication *Advertising Age* observed:

> Frustrated by what they evidently feel is lack of response from the networks and individual broadcasters, the viewers writing the FCC about violence and what they consider obscenity on TV are for the first time mentioning sponsors in significant numbers. Promises of sponsor product boycotts by some individuals and community groups, plus strong indications by others that sponsors are being held accountable to some degree, were included in thousands of protest letters received by FCC over the past ten weeks. About 7,400 were received in February, an increase of 4,600 over the previous month, and the FCC officials see the trend continuing through March and April.[6]

Complaints were beginning to come from organized groups with connections, clout, and a sophisticated understanding of media institutions. In September 1975, *Advertising Age* reported that "an organization called Morality in Media" had presented FCC chairman Richard Wiley with a 100,000-name petition calling for FCC hearings throughout the U.S. on sex and violence issues. Membership of Morality in Media included such high-level executives as retired J. Walter Thompson president Henry Schachte and IT&T president, Francis Dunleavy. Since the FCC indicated little willingness to hold hearings, the article warned, "one of [Morality in Media's] next stops will be Madison Avenue. . . . The organization's 45,000 members will soon be sent the names of the top 10 advertisers and asked to write the presidents of the companies urging them not to advertise on programs with 'gratuitous' sex and violence." Not since the blacklist had large-scale pressure been placed directly on advertisers.[7]

The shift in the 1960s from sole sponsorship to spot advertising had disengaged advertisers from direct program content control. Because commercials for their products appeared throughout the broadcast schedule, it was difficult for companies to be identified with any one program. By eliminating the direct association between product and program, advertisers became less vulnerable to pressure. Standards and practices departments further insulated advertisers by assuming the responsibility for dealing with advocacy groups.[8]

Other mechanisms within the television industry also protected advertisers from financial loss due to pressure groups. Groups were prevented from applying pressure to specific advertisers by the network policy of withholding the names of companies scheduled to advertise in the upcoming program. If advertisers wanted to avoid association with a controversial show, they could simply pull their ads and buy time on the other programs. When viewers wrote protest letters to advertisers after a broadcast, letters of response like the following effectively set the record straight as to who was responsible for program content:

> Our company is not sponsoring any specific programs on television. Rather, our network commercials are purchased on a "scatter" basis. "Scatter" is an industry term used to describe scheduling which is rotated through a number of shows preselected by program title rather than content. For instance, our commercial might be scheduled in *Marcus Welby*, but we would not know whether the show content involves heart trouble, diabetes, or abortion.[9]

While removed from the responsibility and associated headaches of day-to-day content control, advertisers still set the agenda for television programming. Since advertiser dollars paid for all of commercial television, advertiser needs profoundly shaped both the form and content of its programs. When these needs shifted, programming was altered accordingly. When advertisers had decided they wanted to reach younger, urban audiences, television executives and producers willingly complied, and the face of prime time changed dramatically. As one executive explained, "when we started buying demographic audiences instead of households, the networks turned around and cancelled high rated shows that had rural skews because a rating point in programs like *Beverly Hillbillies* was not worth as much as rating points in programs that appealed to more upscale audiences."[10]

It had been many years since organized groups had effectively held advertisers accountable for programming content. In 1975, as violence became a public issue, advertisers sensed their potential vulnerability. Despite the institutional shifts in the industry, advertising research showed that a fair number of people still believed advertisers were directly involved in programming. "As long as three in ten think that advertisers and agencies are responsible for program content," warned one advertising executive, "we will not be able to sit on the sidelines."[11]

Leaders in the advertising industry suggested steps that could be taken to protect their interests. General Food's Media Service Director, Archa O. Knowlton, whose company spent $35 million a year in prime-time spots, urged his fellow advertisers to take unified action to stay out of violent programs "because dollars speak louder than people." "It's time to stop the dialog about whether broadcast violence has a negative impact on society," the executive declared. "Let's be on the safe side and speculate that it can't do anybody any good. We know that violent crimes depicted on television are very often experienced in real life within the next several days. There is reason to believe that violence is contagious and that depicting crime or publicizing crime can stimulate criminal minds to follow suit."[12]

Social responsibility was certainly not the sole motivation for such action. There were other, economic reasons for pulling out of violent shows. The family hour had pushed all the "cops and robbers" shows out of the 8–9 p.m. time period and into the 9–11 slot. This "overscheduling" of violent programs was not necessarily working to the advertisers' advantage. Violent shows were not delivering the right audiences. In the words of one executive: "Current audience composition data show that this program category [cops and robbers] no longer is the surest way to reach young adult women [the most lucrative viewing audience]." Money could be spent more effectively and safely in less violent programs which also attracted the people advertisers most wanted to reach. Collective action might effect substantial changes in programming. "If enough of us were to buy around violent programming," Knowlton told his colleagues, "it would not make business sense for the networks to put such programs or as many such programs on their schedules. . . . If we could get to the point where a rating point in a violent program was worth less to the networks than one in a situation comedy, for example, then maybe the networks would be willing to take the chance of scheduling programming with slightly more downside risk because it would be in their economic interest to do so."[13]

By late 1975, the stage was set for a strong push from organized groups to force a reduction in TV violence. Hearings in Congress and press coverage had put the issue on the public agenda. Complaints were mounting from groups and individuals. And now, most important, the advertising industry had already begun to respond. The issue of television violence galvanized a number of national organizations with separate constituencies and agendas. Some of these groups acted independently; others coordinated their activities. Though there were distinct differences among the individual organizations, their strategies complemented one another. Their combined efforts, and the press coverage they generated, made these groups a strong collective force, capable of bringing substantial pressure to bear upon the television industry.

The National Citizens Committee for Broadcasting was one of the first to take action. NCCB had been concerned about this issue since its beginnings in 1969. Already an active leader in the media reform movement, NCCB knew from experience which strategies could work in producing changes in broadcasting. Like the other media reform groups, NCCB had previously focused most of its efforts on local television stations, whose licenses were the primary pressure point. But the issue of excessive television violence could not be solved by going to individual stations. Nor could effective changes be made by appealing to standards and practices departments. Violence was not a circumscribed, manageable "special interest" issue whose representation could be negotiated. It pervaded the prime-time schedule.

In 1974, Nicholas Johnson took over the leadership of NCCB. He and his colleagues tried to come up with a strategy. As Johnson remembers, "I looked at the past history of violence in television. I had testified many times before government committees. But we saw little change. It seemed there was nothing anyone could do. The networks promised changes, but they didn't follow through." The most effective pressure point for this issue, it seemed, was the advertiser. Since advertisers were already responding to generalized pressure, NCCB leaders looked for a way to intensify the pressure and make *individual* advertisers accountable for program content. To do that, the group borrowed a system that had been devised at the University of Washington. Using social science research methods, a team of researchers had conducted a monitoring study of prime-time programs to identify which companies had bought time in violent programs. Companies were then ranked according to how much TV violence their advertising dollars supported.[14]

Part of NCCB's strategy, Johnson recalls, was to make this monitoring system as legitimate as possible. NCCB looked for top people in the academic and business world to conduct the monitoring study. For a method of measuring violence in programming, the group went to George Gerbner, dean of the prestigious Annenberg School of Communications at the University of Pennsylvania. Since the late sixties, Gerbner and his associates had been conducting yearly surveys of prime-time violence. These reports, called "violence profiles," became important benchmarks for policymakers. Gerbner himself had testified numerous times before governmental committees. NCCB asked Gerbner's staff to train a group of professionals to monitor programming.[15]

"We found a firm that was in the business of screening television for advertisers, to double check that ads had been placed correctly," Johnson remembers. BI Associates was chosen because it was, in Johnson's words, "a conservative, business organization, part of the broadcasting community." The system for monitoring also required the use of a computer, so NCCB selected the biggest computer company in the business, IBM, to provide those services.[16]

By using computer technology and sophisticated scientific methods, NCCB was able to identify the companies that paid for TV violence. Suddenly, the advertisers that for so long had evaded responsibility for programming were publicly exposed. NCCB's first study was released in July of 1976. At a press conference announcing the results, Johnson dramatically illustrated sponsors' culpability for program violence. Television programs, he said, were essentially salesmen for the sponsors. "If a salesman broke into your house by violent means, he'd be put in jail." The six-week study singled out ABC's *S.W.A.T.* as the most violent program on the network airwaves. And the products associated with the most violence in programming were, in order: Tegrin, Burger King, Clorox, Colgate-Palmolive, Gillette, Breck, Ford, Johnson & Johnson, American Motors, and Lysol products.[17]

The NCCB leaders made no mention of a product boycott. Instead, they offered to provide their monitoring program as an "ongoing service" to the advertising industry. If advertisers were willing to pay for the rating services of the A. C. Nielsen company, they could also buy the professional rating services of an advocacy group. As *Broadcasting* magazine explained the plan: "NCCB has developed a rate card under which advertisers, agencies, and others may order various computer-based profiles, analyses and summaries at prices ranging from $100 for a multiweek summary ranking to $2,000 a week for a 'delineated violence commercial profile.' "[18]

Though advertisers were charged a fee for such series, this was not a profit-making venture. The original study had been funded through a grant from a small foundation in California. The ongoing effort received support from additional sources. One of the major funders was the American Medical Association, whose members had spoken out on the TV violence issue earlier. In the December 8, 1975, issue of the *Journal of the American Medical Association,* the organization had printed an article reviewing the research on violence and television and calling the volume of violence on TV a "national scandal." The following year, the AMA announced it would give NCCB $25,000 to help pay for the continuing study.[19]

While NCCB was involved in its campaign, other mainstream organizations were taking on the issue of television violence. One of them was the National Congress of Parents and Teachers, more commonly known as the PTA. Since its establishment in 1897, the PTA had periodically taken up the issue of mass media and children. As early as 1916 the organization had called for a "wise, effective method of censoring motion pictures." Throughout the years the issue of media influence on the young had surfaced in the organization's yearly gathering. Given the extent of the public debate and controversy over television violence, it is not surprising that the delegates to the 1975 national PTA convention took up the anti-violence cause. In August of 1976, declaring that "the public is fed up with violent TV programs," they announced the beginning of a yearlong "war" against television violence.[20]

This declaration added a dramatic new element to the public battle over television violence. With its more than 6,000,000 nationwide members, the organization could hardly be regarded as a narrow special interest group. Virtually every community in America had at least one PTA chapter to look after the needs of schoolchildren. PTA members were angry, and their leaders did not hesitate to use militant rhetoric. The organization had a clearly worked out plan for mobilizing its constituency to action. Using the services of media consultant William Young, the PTA informed the press that it had set aside $110,000 for an action plan that would include public hearings, a grassroots monitoring campaign, license challenges, and sponsor boycotts. While NCCB had made the first move by identifying guilty sponsors, the PTA took the effort one step further by threatening organized consumer action.

PTA leaders set up a private meeting at their Chicago headquarters and invited NBC, CBS, and ABC to send representatives. The networks willingly complied. "It is the threat of the TV networks and the PTA engaged in open warfare," observed *Chicago Tribune* reporter Richard Cheverton, "that drew network executives to the PTA seminar like moths to a dangerous flame." The network delegates tried unsuccessfully to dissuade the PTA from proceeding with its plans. ABC standards and practices vice president Alfred Schneider came armed with various studies showing that there was no causal relationship between violent programming and assaultive behavior. He also assured the gathering that his network had already reduced violence substantially that year. But the PTA went ahead with its campaign. Next, the group planned a series of public hearings around the country, designed to generate wide media coverage.[21]

The hearings were held between November 1976 and March 1977 in eight major cities, including Los Angeles, Chicago, Portland, and Atlanta. Operating like a quasigovernmental agency, PTA leaders interrogated network and advertising executives. Over 500 parents, children, educators, and psychologists testified. The hearings drew considerable press attention and were well attended by the public.[22]

While the debate over TV violence continued to escalate, the advertising industry stepped up its efforts at self-protection. It didn't matter what social scientists had to say about the impact of TV violence on aggressive behavior. Advertisers cared about only one thing: were they going to lose money? To find out, they conducted their own market research. The J. Walter Thompson Company did a study of the effects of the pressure campaign on consumer behavior. The results were alarming. Ten percent of the consumers surveyed considered not buying a product because it was advertised in violent programming, and 8 percent said they actually had not bought a product for that reason. Two out of five said they avoided overly violent programs, while a fifth of the men and a third of the women said they prevented their children from watching such shows. "Aversion to this kind of television is growing," the company's president warned, "and consumers are organizing. This could lead to letters of protest and product boycotts. . . . We are counseling our clients to evaluate the potential negatives of placing commercials in programming perceived to be violent."[23]

Advertisers also began to test the more subtle effects of violent content. There had been interest for some time in the influence of "programming environment" on the effectiveness of commercials. As market research became more sophisticated, advertisers were better able to determine which "environments" made viewers more receptive to the commercial messages. "It is entirely possible," one executive suggested, "that a commercial will work harder in a program that reflects positive social interaction as opposed to one dealing with blood and guts." The J. Walter Thompson Company announced a study to test that hypothesis. "We long ago concluded that there were certain products that didn't live well in surroundings of blood and broken heads. But we have started to wonder about the effects this has on any message."[24]

A few advertisers already had written policies which spelled out the appropriate or inappropriate environments for their commercial messages. Since its early days in radio, Procter & Gamble had limited sponsorship to "programs which are done in good taste, appeal to the people most likely to buy and use our products, and provide a satisfactory setting for our products' advertising messages." Kemper Insurance flatly forbade sponsorship of shows that would encourage reckless driving, such as "programs devoted exclusively to auto, motorcycle or power boat racing."[25]

As violent programs began to be regarded as an unsuitable advertising environment, a number of companies either revised old policies or came up with new ones. Institutionalized guidelines, procedures, and mechanisms were developed to protect advertisers from having their products associated with violent programs and to show the public that they are unwilling to support televised violence.

These companies were not acting unilaterally. While NCCB, the AMA, and the PTA were exposing the sponsors of violent programming, other organizations were engaged in a campaign to get advertisers to develop and enforce policies against violence. A key group involved in this effort was the New York-based Interfaith Center on Corporate Responsibility, which was affiliated with the National Council of Churches. ICCR had been engaged in media activism for some time. Its principal weapon was the stockholder resolution. The organization had recently begun to use stockholder resolutions at the annual meetings of the three major networks to force them to improve their treatment of women and minorities in programs and commercials. In early 1977, ICCR focused its attention on violence and developed a plan to use the stockholder resolutions against the companies named by NCCB in its violence rankings.

ICCR was very successful in its campaign. Most of the companies without policies on violence agreed to adopt them. When ICCR representatives wrote to Kodak proposing that the issue of sponsoring TV violence be taken up on the floor of the upcoming stockholder meeting, management quickly came up with a new policy. "Responding to your letter of December 31 and the shareholder proposal, we have reviewed our policy," wrote the company's legal counsel. "It is not our intent to be associated in our advertising on television with programs which might be considered excessively and gratuitously violent," he continued. "A proposed statement which makes that policy is enclosed. It is our hope that this statement will provide reassurance for the concern expressed in your shareholder resolution."[26]

Kodak agreed not to advertise in television programs which: "1) include violence for its own sake, when violence plays no part in or makes no important contribution to a dramatic statement, 2) contain overly graphic displays of brutality and human suffering, [or] 3) portray anti-social behavior which because of its nature and the manner of portrayal could easily stimulate imitation." Other companies quickly issued new violence guidelines. Among them were Colgate-Palmolive, General Motors, Schlitz, and Pillsbury. Sears Roebuck and McDonald's revised their existing guidelines in response to the requests from ICCR and other organizations.[27]

To implement these policies, advertisers instructed their agencies not to buy into programs which were considered violent. These programs were fairly easy to identify, especially with the help of groups like NCCB. But because there was so much pressure surrounding the violence issue, advertisers wanted further assurances that their spots wouldn't find their way into objectionable programs. To safeguard their commercials against a damaging program "environment," advertisers employed the services of special screening companies. These companies were already part of routine procedures advertisers used in the placement of their commercials. They had come into being in the late fifties and early sixties when advertisers ceased their sponsorship of entire shows. Their primary function was to prescreen network programs to ensure that commercials did not appear in the middle of program content that might be embarrassing or in any way undermine the effect of the commercial message. Any glaring problems would be flagged in advance so commercial spots could be pulled out before a broadcast. The procedure guaranteed, for example, that an airline company commercial wouldn't be shown in a TV program that included an air crash. If there were any question that an advertiser would be at risk, the screening companies would refer the matter to the advertising agency, and to be on the safe side, the agency would exercise the client's option to withdraw the ad. All of this was done within a few days of the broadcast.[28]

This was the first time in the history of television that advertisers had collectively shifted their expenditures based on programming content. *Broadcasting* magazine observed in February 1977, that "a rising chorus of advertisers publicly disassociating themselves from 'violent' television programming in the last few weeks finds many broadcasters apparently convinced it will lead to much less TV violence." Although network executives questioned by the trade publication refused to offer details, they indicated that "there was 'some activity' among advertisers in moving out of some programs into others." Such trends were bound to affect program content.[29]

The networks responded to the pressure campaign in several ways. Privately, they began to look closely at their own programming to see what changes could be made to satisfy advertiser demands. Publicly, they used their in-house social scientists to wage a defensive information war with their critics. Each network had a social research department, whose staff conducted research and commissioned studies from outside scholars on the impact of television. Some of these studies served as counter-research to the more critical work of independent social science researchers. Social research departments performed a number of other functions as well, including: providing standards and practices department with research to aid in their decision making, critiquing the monitoring studies of protesting advocacy groups, and preparing reports for testimony to government bodies.

During the anti-violence campaign, social research departments were particularly important. The House Subcommittee on Communications continued to hold hearings on violence. Professor George Gerbner appeared before the subcommitte in February 1977, reporting that TV violence had increased in the previous fall's schedule. To counter Gerbner's research, as well as the NCCB monitoring studies, the networks began to develop their own monitoring systems. ABC announced to its affiliates that it had devised its own in-house violence index that was much more realistic than the one used by Gerbner. Unlike the "mechanical system" used by Gerbner and NCCB, ABC's "new approach makes distinctions among different types of violence—between comedic and brutal violence." CBS also announced the results of its monitoring project which, contrary to Gerbner's report, showed CBS's violence declining.[30]

Though network efforts to discredit the scientific research may have succeeded in clouding the government investigations, they failed to pacify the pressure groups. In fact, the pressures rose even higher. The AMA announced on April 4 that it planned to step up its anti-violence activities in the coming months, asserting to the press, "We're in this for the long haul." On April 11, the United Church of Christ, which had been responsible for the landmark WLBT decision, announced it too was joining the anti-violence forces. And the PTA made its own dramatic announcement. Having finished their public hearings, PTA leaders announced that their next move was to put the networks "on probation" until the end of the year. During that time the organization would monitor network programming to see whether the networks had reduced violence on entertainment programs. If there were no substantial progress, the PTA would use an arsenal of weapons—including petitions to deny, civil suits, and boycotts—to force the networks to make changes. For the remainder of the year, the organization was going to engage in a massive letter-writing campaign to local station executives.[31]

The centerpiece of the PTA's "action plan" was an elaborate grassroots monitoring project. Its primary purpose was not to provide scientific evidence of programming trends but to mobilize the public. Well publicized in the press, and set in motion at the most intense moment of the anti-violence efforts, the PTA campaign harnessed the rising tide of public dissatisfaction. It also gave citizens a new sense of power. Hundreds of volunteers from around the country were enlisted for the war on television violence. They were trained in two- to three-hour workshops and given a form to fill out as they watched television with their families. They were asked to identify programs that contained violence and to make note of such things as who the initiators of violent acts were ("good guys" or "bad guys"), whether the consequences of violence were shown and if they were realistic, whether violence was used to solve conflicts and whether violent episodes were necessary for the stories' development. Individual violent acts were tallied, and advertisers before, during, and after each program were identified. The volunteers were also asked to rate such things as "quality of life" in the programs, "sexploitation," and stereotyping. "Be your own TV critic," the monitoring form suggested. "Using a program evaluation sheet is helpful in analyzing the programs you and your family view. The results may surprise you. . . . Express your reactions in a letter to your local broadcaster, with copies to the network, the FCC, the advertisers and TV Action Center." Following the NCCB model, the PTA plan called for publication of the names of those advertisers that had bought time in these shows, followed by a "selected product usage" campaign if advertisers continued to support the programs.[32]

By the fall of 1977—after an unprecedented year of pressure from organized groups, the government, and the public—the prime-time schedule had changed dramatically. The networks canceled a number of series targeted by the advocacy groups, including crime dramas such as *The Streets of San Francisco.* Geoffrey Cowan notes, "Only two hard-action police shows,

Baretta and *Starsky and Hutch,* remained on ABC, and not a single new hard-action police program was scheduled by any of the three networks for the 1977–78 season." The networks had also intensified their own internal screening processes using their standards and practices departments. As *TV Guide* observed in August 1977, "There is little violence in next season's schedule, not only because the networks refused to buy new programs that might feature violence, but because returning police and crime shows are being meticulously deviolenced by network standards-and-practices departments." An NCCB news release claimed victory, noting in September 1977 that the violence level was down 5 percent compared with the preceding fall. Even ABC's *Starsky and Hutch*—earlier tagged the "number one show in violence"—averaged only half the violence level of last fall's most violent show. "It looks like the public is at last being heard," remarked NCCB's Nicholas Johnson.[33]

But while violence and hard-action shows were down, there was a decided increase in the numbers of programs that revolved around sex. The networks had warned PTA leaders that "if violence went down, sex would go up." And this prediction seemed to be coming true. Critics labeled 1977 the "T & A" season, a term which *TV Guide* euphemistically translated into "bosoms and buttocks." One of the season's most notable newcomers was ABC's *Three's Company,* which featured three singles living together. Though they weren't sexually involved with each other, misperceptions by others that they were became one of the main sources of the show's humor.[34]

These new trends were particularly troubling to conservative groups. To many of them, the sudden increase in sexual content was a step further into immoral and dangerous programming territory. A number of church groups had already been mobilized as part of the anti-violence effort. With their organizational machinery well oiled for protest, they could quickly launch letter-writing campaigns against other kinds of offensive programs. In addition to flooding ABC with angry letters over *Soap* (see Chapter 5), conservative church groups launched a protest over the NBC TV movie *Jesus of Nazareth* for its depiction of Jesus Christ as a fallible human being. In a departure from standard procedures, the film was entirely sponsored by

General Motors. At a time when advertisers were already feeling intense heat, protesting groups threatened a boycott of GM, which, in turn, withdrew its sponsorship. Procter & Gamble then agreed to sponsor the movie, though for much less money than General Motors had planned to pay for the six-hour series.[35]

In February of 1978, the NCCB announced the results of its most recent thirteen-week study: "Nine out of the top twelve sponsors of hard-action television shows in 1976 greatly reduced their advertising support of 'aggressive personal violence' in television programming during the 1977 season, and the networks decreased portrayal of it by 9% in prime time." Since it looked like their campaign had permanently changed network and advertiser behavior, NCCB and the AMA soon ceased their monitoring of prime-time violence. With content policies firmly in place, advertisers appeared to have stopped their financial support of TV violence altogether. "Like the Vietnam War," Nicholas Johnson remembers, "we declared victory and went home."[36]

The PTA continued its monitoring efforts for several years thereafter, shifting its emphasis away from violence and toward program quality. Monitoring results were periodically released in the form of lists of television's "best and worst" prime-time shows. But that program was also discontinued. In its place, the PTA began two new projects, which reflected a change from the negative, confrontational stance to a more positive, cooperative one, "aimed at promoting 'wholesome' family shows and teaching children to be more critical TV viewers." Observed TV critic Ron Aldridge: "The PTA, which in the 1970s led a nationwide assault on video violence, is getting involved in television again. This time, however, the emphasis is on coping with the tube, not reforming it." To help members "cope" with television, the PTA began a program of teaching "critical viewing skills" to its grassroots organizations. It also set up a "TV review panel" based in Los Angeles. The idea for such a panel came from discussions between PTA leaders and TV industry representatives. Its function was to prescreen programs and to issue endorsements which could be used by the industry in its promotion of the program. The panel charged a fee of

$250 to cover expenses, and evaluated the material on the basis of such criteria as social value, violence, profanity, sexual material, stereotyping, plot, and production quality."[37]

Why did the PTA decide to change its strategy toward network television? Like NCCB, the PTA leaders believed that its work was done. "We did see a significant decrease in gratuitous violence," one representative reported. PTA leaders also found that an intense public campaign simply could not be sustained for an indefinite period of time. Leadership shifted, members began to tire of the issue, and the organization readjusted its priorities, shifting its attention to other issues such as drug abuse and alcoholism. Noted Warren Ashley, who served as a consultant to the PTA, "Many members began to worry that maybe the kind of publicity the PTA was getting was bad for its image. We felt the PTA needed to take a more positive approach, a more constructive approach." After his involvement with the PTA campaign, Ashley went to work for NBC as a standards and practices editor.[38]

Finally, PTA members believed that the entry of such new media as satellite and cable television would change the structure of the television industry, making the networks less important. "We had been arguing for alternative choices," one leader explained, "and now with the new technologies, a multiplicity of choice was upon us. We realized monitoring was not necessary, a better strategy was to train people to use television." As a PTA newsletter informed its members: "It is no longer possible to improve television merely by persuading the major networks to get their houses in order. Television is exploding into hundreds of channels and shows on dozens of networks. The only place to deal with it is at the consumer level."[39]

The campaign to reduce prime-time violence was the most intense, broadbased effort in the history of television. The groups that participated in the campaign had come up with a unique strategy for making advertisers vulnerable to pressure and bringing them back into a more direct relationship with programming content. The campaign produced dramatic effects in prime-time programming within a fairly short period of time. It also caused new mechanisms of control to be established within the advertising community and the network institutions. Nevertheless, its impact on violent content was short-lived. When the pressure subsided, and advertisers became less vulnerable, levels of violence began to rise again.[40]

Other organizations tried to pick up where the PTA, NCCB, and the AMA left off. A group of individuals who had worked with NCCB formed the National Coalition on Television Violence. Headed by Dr. Thomas Radecki, a psychiatrist based in Illinois, NCTV continued to monitor violent content in all media, periodically releasing the results to the press. But NCTV had neither the constituency, the press attention, nor the historical moment that had contributed to the success of the earlier campaign. NCTV was not considered a serious threat by the networks.[41]

Although the anti-violence campaign may have had only a temporary effect on prime-time content, it left its impact on the television industry. Advertisers learned an important lesson about their potential vulnerability to pressure and developed strategies to better protect themselves. After 1977, there were mechanisms in place for more careful routinized surveillance of the programming in which their commercial messages appeared. The content policies, and the screening companies that carried them out, were able to respond quickly to the winds of pressure and to flag sensitive content. And advertisers also continued to conduct research on the relationship between the programming "environment" and the effectiveness of commercials. Such research could give advertisers much more sophisticated tools for knowing exactly what kinds of programming content would work best for their commercials and which kinds should be avoided. Similarly, the networks had developed their own sophisticated research techniques for tracking sensitive content. These would be fed back into the content policy area to help networks handle violent content as they had learned and were continuing to learn to handle other kinds of sensitive program material.[42]

For the organized groups outside the industry, the experience with the violence issue had demonstrated that a well-orchestrated campaign could get the television industry to respond.

When the International Association of Machinists decided in 1980 to launch a campaign to improve the image of labor in prime-time television, it hired William Young, the consultant who had developed the PTA campaign. IAM then set up an elaborate grassroots monitoring program on the PTA model and released its findings to the press, which showed that the American worker was significantly under-represented in prime-time. Though the organization threatened tough action to pressure the networks into improving its image of workers, very little came of this campaign. The main reason was that this strategy of grassroots monitoring and public exposure wasn't really suited to the issue that concerned IAM. Although the campaign probably served to educate rank and file union members about the structure of television and helped them to develop critical viewing skills, it failed to enlist the support of the American public.[43]

It would not be long, however, before the strategy devised by anti-violence forces would be used again.

CHAPTER SEVEN

Battle over *Beulah Land*

It was a Hollywood event. About 350 black TV and film personalities were gathered at the Mark Taper Forum, a small theater next door to the Dorothy Chandler Pavillion, where the Academy Awards ceremony was held each year. On this cool February evening in 1980, almost everyone who was anyone in the Hollywood black community was there—Sidney Poitier, Esther Rolle, Ossie Davis—along with lesser-known black actors, writers, producers, and directors.

But this was not a social gathering. It was the first meeting of the Media Forum, a newly formed black activist group. Funded by the Brotherhood Crusade, the Media Forum was organized to bring public attention to the status of blacks in the entertainment industry. Focused on such issues as employment, media ownership, and elimination of stereotypes, the group's primary strategy was to hold a series of public meetings with leaders from government and industry. This was the first such meeting, and Charles Ferris, chairman of the Federal Communications Commission, had been invited to be the featured speaker.

Ferris was the only white person seated among the black reporters and dignitaries on the stage. The scenery for the Neil Simon play *I Oughta Be in Pictures* was still in place. With his pale complexion, and thick, snow-white hair, Ferris looked strangely out of place. Those gathered that evening listened closely to what the chairman had to tell them. Though blacks had made some gains in television, many were disappointed at the limitations on their progress. Appeals to the FCC had helped blacks in the past. Maybe the regulatory agency could be a source of power again. Black activists approached this first meeting with hope.

But Ferris's words were anything but encouraging. There really wasn't much the FCC could do to eliminate stereotypes on TV or to increase roles for blacks, he told the group. One must never forget that network television is a business, explained the commissioner. "Solutions must be based on the full awareness of the cold economic realities of the electronic media business." But he offered no such solutions, and as his speech continued, one could sense the growing frustration in the room.

The extent of the audience's displeasure became more evident when actor-director Ivan Dixon rose to the podium, pointed to Ferris and delivered an angry response: "He said essentially nothing to help us in any way to crack the system. What this man said is that 20 years from now your children will be sitting right here asking questions of some other white chairman of the FCC. . . . The only thing to do, Dixon told the audience, was to "get of your asses and light a fire under *his* [Ferris's]!" Dixon's remarks were met with cheers and applause. "And another thing," he added, "we are not going to allow the networks to force this *Beulah Land* on us!" The words *"Beulah Land"* were like a war cry to the crowd. An explosion of angry shouting burst out immediately. The upcoming miniseries had become a symbol and a rallying point for the growing frustration and rage among the Hollywood black community.[1]

Within weeks the battle over *Beulah Land* was to become the most dramatic confrontation between black Americans and network television since the *Amos 'n' Andy* controversy.

The NBC network, Columbia Pictures, and producer David Gerber had not anticipated the protest that erupted over this adaptation of the bestselling novel by Lonnie Coleman. When the development began on the project in 1978, *Beulah Land* was conceived as a *"Gone with the Wind* for television." Like the 1939 film classic, *Beulah Land* was a "romantic Civil War saga." As prime-time fare, it was also consciously modeled after *Roots*, the successful 1977 ABC miniseries which traced the story of a black family through several generations. Gerber characterized

Beulah Land as "the flip side of *Roots* . . . showing whites going through those same years instead of blacks."[2]

Gerber was under no illusion that *Beulah Land* measured up to the quality of either *Roots* or *Gone with the Wind*, but it seemed like a sure winner for prime time, with all the necessary elements to attract TV's most valued demographic group, women 18 to 49 years old. Referring to it as a "potboiler," the producer described the show as "a very romantic story . . . a romantic novel . . . soap opera . . . a gothic romance with a plantation instead of a castle."[3]

NBC president Fred Silverman—who had moved from his job as ABC's programming chief—must have thought *Beulah Land* was perfect TV material too. He selected the project for development as a six-hour miniseries scheduled for the May ratings sweep period.* With a budget of $10 million, *Beulah Land* was to be NBC's costliest production.[4]

If the script for *Beulah Land* had been submitted to the standards and practices department in the early seventies, it would have raised a red flag and very likely not have been approved, for fear of protests from black activists. But times had changed. Most of the political activism of the media reform era had died down. The militant advocacy groups that had exerted such pressure on the networks a few years before were no longer campaigning against network television. Many of them had disappeared. Broadcast industry lobbying in Washington had begun to have an impact, and "deregulation" was already under way. The petition to deny, which in the early days had served as a powerful Damocles sword, had lost much of its effectiveness, since the FCC seldom ruled in favor of the citizens' groups. Many minority groups had begun to believe that it wasn't worth the cost and trouble to file them anymore.[5]

A few advocacy groups—like the gay activists and the Gray Panthers—had worked out institutionalized relationships with network standards and practices departments, which gave them some input into networking programming. Black groups, however, had not established such a relationship. The NAACP had

*National ratings sweep periods occur four times a year, when the ratings services measure all the TV markets around the country. The results are used to set the advertising rates for the following quarter.

maintained an office in Hollywood, but much of the group's efforts had been directed at increasing employment, rather than negotiating the portrayal of blacks in entertainment television. The organization had shifted its one-time confrontational stance to a much more conciliatory one, presenting yearly "Image" awards to members of the film and television industry for their positive treatment of blacks. Gerber himself had received one.[6]

But network programs have long gestation periods before they appear in the prime-time schedule. As this new miniseries was working its way through the development process, new events were coalescing to create the environment that was to transform *Beulah Land* from just another mediocre TV show into the center of a large-scale battle.

The success of *Roots* in 1977 and its sequel, *Roots: The Next Generation*, two years later, had seemed a good omen to blacks. It looked like TV had entered a new era where blacks and black themes could receive full dramatic treatment and be accepted by a mass public. Black actors were particularly hopeful that new opportunities for major roles would open up to them. Their optimism was short-lived, however, as new black dramatic TV series failed. Particularly disappointing was the fate of two shows introduced in the 1979 fall prime-time schedule. *The Lazarus Syndrome*, an ABC medical drama starring Lou Gossett, Jr., vanished within a few weeks of its appearance. Its cancellation was followed shortly thereafter by the demise of *Paris*, a CBS detective show with James Earl Jones in the title role. Neither show had garnered sufficient ratings, and word around Hollywood was that the public didn't want black drama on TV.[7]

Many blacks in the industry were convinced that renewed political activism was necessary. Groups including Concerned Black Artists for Action, Actors Speak for Life, and the League of Black Cinema Artists started working together to improve their status in the TV industry. They were heartened by the recent cancellation of *Mr. Dugan*, a comedy series about a black congressman. It had been withdrawn a few days before its scheduled debut as a result of protests by black elected officials. Its executive producer, Norman Lear, had unilaterally made the decision to kill it despite the strong objections of CBS network executives. Though a black political consultant had been employed during development of the series, Lear, who had not supervised production of the pilot, was both disappointed and concerned about its final form. As he remembered the pilot:

> Instead of being about a Congressman who was straight and effective and amusing, the lead actor played it like a buffoon. And it just wasn't right to tag a black—the first time you were going to do a black elected official—to make him such a fool.[8]

Lear flew to Washington and pre-screened the pilot episode before the seventeen-member Congressional Black Caucus. Caucus members were so upset and offended by what they saw that many of them stormed out of the room in the middle of the program. An embarrassed Lear promised them that he would never allow it to air. Rather than alienate black political leaders, Lear's production company took a loss of more than three-quarters of a million dollars by cancelling its contract for the series.[9]

This was a very different case from *Beulah Land. Mr. Dugan* was a contemporary situation comedy about a specific group of powerful people, and it was that group that strongly objected to the series. But despite these major differences, both the timing and the circumstances of *Mr. Dugan*'s withdrawal awakened a new spirit of power among blacks in Hollywood.[10]

It was against this backdrop that casting began for *Beulah Land* in the late fall 1979.

The conflict between the need to find work and the humiliation of accepting demeaning roles has often created difficult choices for minority actors. This had been an issue during the *Amos 'n' Andy* controversy and it resurfaced now. *Beulah Land* offered many jobs for black actors, but it required them to make painful decisions. For some, there was no question about what to do. Said Reuben Cannon, a major black casting director in Hollywood: "When I got to page 9, I knew I couldn't be part of it. It's the most racist piece of material I've ever read. It makes *Mandingo* look like a classic."[11]

Several major actors who found the script offensive accepted parts in the movie anyway, believing they could negotiate changes in their roles during the filming. "I'm not offering excuses," explained Dorian Harewood, who played Simon Haley in *Roots: The Next Generation,* "but I was offered so much crap after *Roots.* . . . It got to the point economically where I had to accept the next thing that was offered. And the part of Floyd—who becomes overseer of the plantation in *Beulah Land*—is a good one." Harewood didn't get a chance to read the whole screenplay until he was already on a plane to Natchez (Mississippi) where the film was to be shot. By the time Harewood and his wife, actress Ann McCurry, got off the plane, they had finished reading it, and it made them sick.[12]

Harewood stayed with the project anyway, hoping he could "create some insight into this character." He justified his decision with the rationale that his stature as an actor, the cooperative attitudes of the line producer and the director, and the fact that there were a lot of black people in the production would give him some power to make changes in the script. "Hundreds of actors would love to play this role, but I felt I would do things with this character and fight for things another actor might not fight for."[13]

James McEachin, who had starred in the NBC detective series *Tenafly,* also found the script offensive but he too believed he might be able to make some positive changes during the production. "I accepted the role, number one, because I needed the money, and number two, because of the character's inherent integrity. Since I did not do *Roots,* I thought here was the opportunity for me, as an actor, to say something about slavery."[14]

When other groups had tried to protest television programs that were still in production, they often had difficulty getting access to scripts. Gays used their own spies to smuggle the *Marcus Welby* script out of ABC before launching their campaign against that network. Other groups relied on rumor and word of mouth to determine whether a program yet to be broadcast was offensive enough to rally support against it. This was not the case with *Beulah Land.* Because so many black actors were participating in the casting process, the script was very accessible. A number of actors who read for parts were so upset that they decided to show the script to some of their friends and colleagues. At precisely the moment when frustration among Hollywood blacks was at a peak, the *Beulah Land* script provided a concrete object for anger and served as a catalyst for the spontaneous mobilization of a protest.

What was it about the *Beulah Land* script that so enraged the black actors? The subject matter itself was part of the problem. Many blacks felt that any program showing them as slaves could damage their recent strides in television. As one of the actor/activists put it: "*Roots* should have closed the book on slave drama. We said it already. We need to get on with living." The particular portrayals of slaves were also part of the problem. The script contained a number of characters and scenes that were stereotypical and inflammatory. According to the activists, "Every one of the approximately fifteen black speaking roles in the script is negative and perpetuates the image of the slave as ignorant, oversexed, sloven, dependent on the whim of his master and filled with love for that master and that master's land."[15]

There were several scenes in the script that became symbols throughout the protest, cited over and over as examples of the insensitivity displayed in *Beulah Land.* One scene showed a slave woman breastfeeding a white child along with her own baby. The stage directions referred to the "wonderful sight" of the two babies. This aroused the ire of the activists, who concluded that "the creators of this show think such behavior is 'wonderful' because it relieves them of guilt and the necessity to deal with the brutality and emotional rape required to force a woman to nurse the children of her enslaver at her own breast." There were objections to another scene, in which slaves were shown participating in and perpetuating their own mistreatment. In it, a black slave boy (Floyd) and a white boy (Leon, later to become master of the plantation) are eating watermelon together. Ezra, the black boy's father, scolds them with these words: "Mah own son! Hepin' the young maisa do bad! Shame on you Floyd! An' shame on you, Leon! 'Specially you! Yo mama waitin' for you to greet the guests like a young gennamun an' you here in this mudhole like a hawg! You gwine be de maisa! You start actin' like the maisa, you hear!"[16]

Beulah Land was an especially easy target, because it was filled with violent and sordid content, and included very little of redeeming value. Those who read it for the first time made laundry lists of the offensive material in the script: "two suicides, four murders on camera (with more occurring off-camera), two rapes, a lesbian relationship that begins when the girls are 12, and the birth of a mulatto baby who is a hunchback."[17]

As the casting continued and preparations were begun for production, the script circulated among black members of the Hollywood creative community. Representatives of the Hollywood Beverly Hills NAACP called David Gerber at Columbia pictures and demanded a meeting. The meeting, which took place in December, left the activists feeling frustrated and the producer feeling satisfied that he had met the objections to *Beulah Land*. The various participants had conflicting recollections of the event. According to Geraldine Green of the NAACP, Gerber's behavior was "evasive." Although he promised a later meeting to discuss the matter, it never came about. When it began to appear that the producer was not going to be very cooperative, Green warned Gerber's office that "we would take very strong action if that program was aired as is." Gerber's recollection of the December meeting was that it was "very conciliatory" and concentrated on "trying to do more with ethnic groups in the overall production." Gerber said he assured the NAACP that writer J. P. Miller had done "a helluva job" on the script and that *Beulah Land* was historically accurate. It was Gerber's understanding that the one meeting resolved the issue and he denied that any further discussion was promised. A subsequent statement by a Columbia Pictures spokesperson indicated that the NAACP had no objections at all to *Beulah Land* during the meeting.[18]

By December 29, casting was completed and filming began on location in Natchez, Mississippi. There, the actors who had hoped to be able to change certain offensive elements of the script found they had little power. In one of the scenes, McEachin's character, Ezra, is set free. Instead of responding gleefully to the news of his freedom, Ezra, who had been a slave on the plantation all of his life, says: "So that means we got's to leave *Beulah Land?*" McEachin objected to that line. Although he could understand how the elderly Ezra might be reluctant to leave his home, "I wanted to at least show the dignity of the man and that a quest for freedom was still there. But you think those goddamned idiots would do that?" In another scene, Ezra watches as the plantation owner beats Ezra's son. "I told them [the director] my conscience would not allow me to do this. It was casting aspersions on 35 million persons by showing the black man has no guts at all." In this case the actor worked out a compromise with the director, where Ezra was allowed to take the beating for his son instead of watching helplessly from the sidelines. "Everyone loved it," according to McEachin. "But the next day, when I was off, they reshot the scene without me, the old way, so the boy gets the beating." This was followed in the script by another scene where the slaves are shown dancing and laughing, with Ezra playing music. "Can you imagine that?" McEachin told reporters, "Playing music and dancing right after my son is beaten? The director told us to laugh it up, but there was nothing funny. After awhile, they beat you down, so I laughed." Having taken the role with hope that he could make changes, McEachin concluded that he had "nothing whatever to say in a favorable light about the asinine piece of garbage called *Beulah Land*. It's denigrating, and an embarrassment. I'm ashamed to admit I played a part in it."[19]

While McEachin battled it out on the set, back in Hollywood a collective effort against the film was gaining momentum. As the script continued to circulate, more and more people believed not only that action should be taken to protest the film but that black people should unite to keep *Beulah Land* off the air. While opposition to the movie was clearcut, a strategy for preventing its airing was not. "We really weren't sure what to do," said Robert Price, who was one of the organizers of the protest. One of the concerns was that too much publicity against the film could backfire, and if they could not keep it off the air, all their efforts might just draw more attention to *Beulah Land* and in fact boost the ratings. Still they felt they had to do something. "We were very cautious. . . . We realized that we could run the risk of sensationalizing the issue. The more attention we point to this, the bigger risk we have. And we also knew that taking on David Gerber, Columbia Studios and NBC was easier said than done."[20]

To expand their power base, the organizers decided to spread the word about *Beulah Land* to political leaders. Continuing to circulate the script seemed like a good plan, but it posed some problems. The main drawback was the sheer size of the document. "We couldn't even afford to get a copy to send to people who should see it," said Price. And besides, even if they could, it was too much to digest. "You can't very well send a 315 page script to a Congressman and say, 'What do you think of that?' I realized somebody was going to have to synopsize the contents so people could have access to it." So Robert Price and another activist, Saundra Sharp, wrote a twenty-three page document entitled "A Position Paper Against the Airing of *Beulah Land*."[21]

Subtitled "A Call to Create, Nurture, and Protect Positive Black Images," the position paper was primarily a series of excerpts of the most inflammatory scenes and dialogue, xeroxed directly from the script, with some interpretation and analysis interspersed. It characterized *Beulah Land* as "intensely offensive and degrading to black people," and called for action against the film. "We must let those responsible know that we have no intention of allowing this assault on our identity to be aired without our organized and concentrated resistance," the paper declared. "We are urging all individuals and organizations to actively oppose the airing of *Beulah Land*." No specific plan of action was included in the document, which was released in February 1980, the same time as the Media Forum meeting was held.[22]

While protest organizers continued to debate the pros and cons of going to the press, Ivan Dixon's unplanned outburst made the decision for them, instantly turning *Beulah Land* into a public issue. The Hollywood trade papers picked up the story. David Cuthbert, reporter for the New Orleans *Times-Picayune*, read the trade pieces on the controversy and decided to do his own story about the trouble brewing on the set of *Beulah Land* in nearby Natchez, Mississippi. "A storm of protest is ready to erupt over the six-hour NBC-TV movie *Beulah Land*," he wrote in an article entitled "TV *Beulah Land* Worse than *Mandingo?*"[23]

Despite the fact that NBC had already spent $10 million on *Beulah Land*, the activists were determined to see to it that *Beu-*

lah Land never reached the airwaves. "We're not concerned with how much *Beulah Land* cost," one of them told the press. "They can afford to flush that money down the toilet." And if it's taken off the air, he added, "maybe they won't be as likely to take such foolish steps in the future."[24]

The day after the first article in the *Times-Picayune,* the director of *Beulah Land,* Virgil Vogel, suffered a heart attack and had to be hospitalized. The production was temporarily halted, and the actors and crew returned to Hollywood. The timing of the director's illness was a critical incident for the protest: it set the production schedule behind for a few weeks; it allowed the dissident members of the cast to return to Los Angeles and participate in the quickly-developing movement to keep *Beulah Land* off the air, and it freed the actors to speak out publicly about their opposition to the miniseries.

By this time the organizers of the protest were fully engaged in publicizing the issue. More groups began to join the protest, coalescing into a "Coalition Against the Airing of *Beulah Land.*" Among the Coalition's members were not only the original four groups but also the Hollywood-Beverly Hills NAACP and two other activist groups—the Association of Asian-Pacific American Artists, and Women Against Violence Against Women.[25]

By the time *Beulah Land* had resumed production on March 3 with a new director, Harry Falk, the controversy was widely known. Not only was the *Times-Picayune* continuing to follow the case but it was being picked up elsewhere in the press. This expanded coverage also brought the first public reactions from producer David Gerber, who labeled the protesters "misguided zealots who are shooting from the hip and spouting off at the mouth." Charging that he was being "tried in the press," Gerber adamantly defended the film, telling reporters, "There are no racial stereotypes in *Beulah Land.* The black people here are slaves. Some are very ignorant. . . . Some of them got their freedom and didn't know what to do with it. Some of them did have that devotion to their masters. It wasn't all hate and brutality." The producer swore he would not let "hysteria warp history." "I don't intend to change a word of the production. I won't bow to emotional zealotry. I'll quit the business first . . . and I'm warning those people. If they do anything to hamper this project or damage my reputation, they could end up facing me in court."[26]

If Gerber was entrenched in his position, so were the leaders of the newly formed Coalition. "*Beulah Land*'s detractors want more than changes in the script," NAACP's Collete Wood told the press. "Our aim is to prevent the airing of the show."[27]

With Gerber publicly attacking the protest, the leaders of the Coalition escalated their efforts, trying every means possible to get the word out that *Beulah Land* must be stopped. "So that we could not be labeled as an isolated disgruntled actors' group," Price recalls, "we began to send the position paper to members of the Congressional Black Caucus. We also contacted [FCC Commissioner] Tyrone Brown. We left a copy of the position paper with Congressman [Augustus] Hawkins's office and that office agreed to copy it and distribute it to the rest of the members of the Caucus. It snowballed. It wasn't anything so strategic as to say we were going to send a thousand to the South, Midwest, etc. It was wherever we had access. . . . You see what happened is that we had people who were so incensed by the debate that they would go to the office and run off 25 additional copies of the position paper and send them to people who they thought should know about this. As a result, this position paper went all over the country."[28]

Since the Coalition had many members who were celebrities, they organized an ad hoc publicity tour for the protest. Using local television community affairs shows—many of which were hosted by blacks—coalition members began making appearances in different cities around the country. If someone were taking a trip anyway, he or she would make arrangements to appear on a program and discuss *Beulah Land*.[29]

On March 7, the Coalition ran a full-page ad in *Daily Variety:* "An open letter . . . re: *Beulah Land* . . . to: David Gerber, David Gerber Productions, Fred Silverman, President, NBC-TV, Larry White, President Television Division, Columbia Pictures and Edgar Griffiths, Chairman of the Board, RCA Corp." The ad read: "The black character images, as depicted in the script of the television mini-series *Beulah Land* are an affront to civilized people everywhere. . . . We feel that the airing of such derogatory images is insensitive, demeaning and dangerous, and tantamount to a direct political attack on blacks and on women, for both groups are portrayed as mindless objects to be used in every vile, degrading manner imaginable! THEREFORE, WE THE UNDERSIGNED URGE NBC-TV AND ITS AFFILIATES TO REFRAIN FROM AIRING THIS AND SIMILAR MATERIAL!" The ad listed the six organizations making up the coalition, and it was signed by 66 people, including actors Robert Hooks, Ivan Dixon, and John Amos.[30]

While the local branch of the NAACP was taking a strong position against *Beulah Land,* the national organization was much more cautious. In a guarded statement on March 2, a spokesperson told the press, "The organization will definitely join the protest, but with 1800 branches and half-million members, we have to be careful how we proceed." At around that same time, however, NAACP president Benjamin Hooks, a former FCC commissioner, had taken direct action by sending telegrams to Fred Silverman, David Gerber, and Larry White, urging them to discuss the film with members of the Coalition. The telegram, which was circulated by the Coalition at about the same time the ad appeared in *Daily Variety,* read:[31]

> The National Office of the NAACP has received numerous phone calls concerning the production of the TV miniseries, *Beulah Land,* scheduled for airing on NBC. Persons whose opinions and judgements I respect have decried the film's depiction of blacks in demeaning, racially derogatory roles in scenes with no historical integrity. While I have not had a chance to review the entire script and am not able to make an independent assessment, I believe that the producers and responsible TV broadcasters should deal with the serious objections being raised about the production. I therefore urge that before proceeding further with the production, that you sit down with the black artists and local NAACP representatives to discuss how the project can more accurately and tastefully deal with the cruelties of slavery and life on a plantation. This subject can be handled with taste and grace. It ought to be. . . . I plan to assign a staff member to look into this fictionalized account of an ugly and barbaric part of the American history so that I might be better able to advise directly in the near future.[32]

The protesters had made no attempt to contact the network directly when they first began their campaign, even though they had been unsuccessful at getting cooperation from the producer. Instead, they concentrated on taking their case to the press in order to rally support. Needless to say, the network would have preferred to keep the whole affair quiet. Jay Rodriguez, NBC's West Coast vice president for corporate information, "flagged it as a potential problem" at the time of the Media Forum meeting in February, which he read about in the trades. "If they'd come to me," explained the executive, "I would have probably gotten into it earlier. . . . I would have gotten Broadcast Standards in very quickly, and said 'Hey, we've got a problem with this program, what's it all about?' We would have had a meeting with Gerber to see what the problems were and whether there were some solutions to the problems. . . . But, they never did contact NBC; NBC contacted them."[33]

Coalition leaders were reluctant to meet with NBC at first, fearing that it would be only an empty gesture and that no one with any power within the network would be there to listen to their grievances. But they finally agreed. On March 11, members of the Coalition met with representatives of NBC and Columbia Pictures for two hours at the West Coast headquarters of NBC. Gerber did not attend the meeting, but several Columbia Pictures executives were there, along with NBC executives from broadcast standards, programming, and public relations. The industry representatives tried to assure the Coalition that the script they had opposed had been revised many times since the actors had first seen it in December. The protesters asked for documentation to back up these assurances, telling the executives, "We feel that it *[Beulah Land]* is derogatory and degrading, and as far as we're concerned, it's not salvageable. If we are in error, we would like very much to see an updated script, any footage or anything else, just to convince us that we are wrong. You know, we'd like very much to be wrong."[34]

Gerber told the press the following day that he had no intentions of giving Coalition members a copy of the script. "That's censorship," he charged. "I'm not about to give a script to someone not involved in the production of the show."[35]

While Gerber publicly refused to give in to pressure, executives at the network quietly negotiated with the protesters. A copy of the script was mailed to the Coalition on March 14, along with a letter assuring them that further changes were being made. "Many changes made are not reflected in this script," the letter read, "and many more will be made as the shooting continues. Since no final document is available at this time, this is the best we have. . . . As the Coalition's position is not only the objection to isolated scenes but also to the totality of *Beulah Land,* we will show you complete segments in rough-cut form (no sound effects or music). We hope to have that rough cut delivered to NBC in approximately two weeks. . . ." The letter also admitted that "the writer's interpretation of the black dialect in the script is understandably offensive" but that "the changes the director and the actors have made in the performance will be apparent when you view the rough cut."[36]

The network did not indicate publicly that it had shared the script with the Coalition. There were already indications in NBC's comments to the press that the miniseries might be postponed. When asked to confirm the date of broadcast, the network spokesman replied: "We have agreed to continue a dialogue [with the Coalition], and until that dialogue is completed we'll have no further comment."[37]

The script which NBC mailed to the Coalition failed to mollify the protesters. In a letter dated March 17, 1980, Robert Price told Jay Rodriguez:

> . . . we do not find your letter responsive to our requests and there is nothing in the "revised script" to assuage our fears regarding its danger to black people and the image the show projects in our society. . . . The revised script is essentially the same script that was the basis for our original opposition. There are no major revisions that encompass our concerns. Revision, editing or reshooting is pointless without a basic sensitivity to the racial implications of this story. There is nothing in this version of the script to indicate that restructuring is the objective of the revision. . . . [T]his revised script does not offer any hope that shooting or editing will make the project less offensive. Therefore, our public opposition to *Beulah Land* continues.[38]

While NBC had been trying to negotiate with the Coalition, both the network and the studio were receiving more pressure from outside in support of the Coalition's position. The press continued to carry the controversy. Several television critics took unusually strong positions. *Los Angeles Times* critic Howard Rosenberg wrote: "My feeling, based on reading the shooting script, is that *Beulah Land* shouldn't be made since TV's record in portraying blacks already has a negative tilt. Except for the *Roots* epics and a few other productions, TV's depictions of blacks have not been praiseworthy. There is little on the books to balance an atypical portrayal that could perpetuate harmful stereotyping. What we don't need is a TV version of *Gone with the Wind.*" The *Kansas City Star*'s TV-radio critic Gerald Jordan, reviewing the controversy in the context of TV's overall treatment of blacks, argued: "On balance . . . *Beulah Land* appears to serve no useful purpose." And many writers in the black press urged their readers to join the protest against the planned broadcast. *Chicago Metro News* urged people to support a "boycott of NBC and its affiliates until they show some semblance of responsibility to its large black viewing audience and stop showing stereotypical movies that only perpetuate falsehoods and serve no good purpose."[39]

Black political leaders who had joined the protest began writing directly to the network executives. Congressman Augustus Hawkins, (D.-Calif.) wrote on March 25 to both NBC and David Gerber:

> I have been receiving numerous communications from my constituents, from responsible local arts and entertainment organizations, and from other citizens groups in Los Angeles, expressing serious concern about the demeaning portrayal of blacks in the *Beulah Land* film production. I am writing to you because I believe your company needs to meet with these organizations; to explore ways of addressing the important issues they are raising. Although I have not read all of the *Beulah Land* script, I have read enough of the script to be concerned about the questionable portrayal of the black condition in the pre-Civil War and post-Civil War periods in the South. . . . It is my opinion that this film should not be aired until these matters are fully addressed and resolved.[40]

The response Hawkins received from the network and the studio assured the congressman that he need not worry about *Beulah Land.* In his letter to Hawkins, Larry White, president of Columbia Pictures TV, told the congressman that his company was "committed to a dialogue which I am sure will alleviate these concerns." He added, however, that such a dialogue could continue only "if they [the Coalition members] will agree to meet with us. Let me assure you," White continued, "that in connection with the production of *Beulah Land* as well as all other programming undertaken by Columbia Pictures Television, we believe we have been continuously sensitive to the concerns of the contemporary black community. By no means is it our intention in any way to degrade or distort the history of the black condition in the pre-Civil War and post-Civil War periods, but only to reflect the full spectrum of life in those times."[41]

In a separate response to Congressman Hawkins, NBC's programming chief Brandon Tartikoff wrote: "We appreciate knowing of your concerns. Similar concerns have been expressed by some blacks in the Hollywood production community, and we have been meeting with this group in ongoing discussions that we hope will be helpful. You may be sure that NBC's decisions in this matter will be made responsibly and with regard for the interests and sensitivities of all segments of our audience."[42]

Despite these assurances of "ongoing dialogue," NBC had reached a definite impasse in its negotiations with the Coalition Against the Airing of *Beulah Land*. The Coalition leaders had made it very clear that they were not satisfied with the changes made so far in the production and there was little to suggest any further changes would assuage their anger. By the end of March, the network had still not made an official decision in response to the Coalition's latest position.

Meanwhile, Gerber and other members of the Hollywood production community were gathering their own forces in opposition to the Coalition and were beginning to make collective public statements, not only about *Beulah Land* but about other pressures from outside groups. Having weathered the recent anti-violence campaigns, all three networks were now involved in confrontations with ethnic groups. Jewish groups were waging a large-scale protest against CBS for casting PLO supporter Vanessa Redgrave in the role of a Jewish concentration camp survivor in *Playing for Time*. And the production of ABC's miniseries *Hanta Yo* was in serious trouble because of opposition from Sioux Indian groups. The Caucus of Producers, Writers, and Directors issued a public statement about the *Beulah Land* controversy in late March formally opposing what it called "prior censorship." "We firmly believe," that statement read, "that, no matter how well-intentioned or how worthy the motivation, it is contrary to every precept of creative freedom for pressure groups to indulge in prior censorship by attempting to intimidate a television network or station from airing any given program." The statement labeled the tactics used to keep programs off the air as "analogous" to "book burning," which "serve no one's cause except those who wish to put further limits on the constitutional and personal liberties to which all of us are entitled." The appropriate remedy for "harm to any person or group" which is "perpetrated or perpetuated" by television, argued the statement, is "through the courts or by way of public opinion."[43]

By the first of April, with the scheduled air date slightly more than a month away, NBC found that it could please neither the protesters nor the producer with its standard techniques of pre-screening and consultation. So the network tried another tactic—postponement. *Beulah Land* was being "rescheduled from May to the fall," NBC announced on April 4, since "production is now at a point where a final print of the complete program will not be available for review until May." This delay, the network said, would allow further time for

> consultation with experts and others interested in the film as produced. We want the program to have entertainment value, to be successful, and we are sensitive to questions about the portrayal of blacks in programs of this kind. The fall scheduling will give us all sufficient time to make sure that the program is a success and sensitive to concerns about the portrayal of blacks. NBC recognizes its own responsibility for its program judgments. In programs set in the Civil War era, NBC makes a particular effort to reflect historical accuracy and a respect for the struggle by blacks for human dignity and freedom.[44]

Coalition leaders regarded NBC's decision as a partial victory, but also viewed it as a tactical move. Said NAACP's Collette Wood, "I don't think they [NBC] are being benign or being kind. It may very well be that they think we'll all disappear by fall." But the Coalition vowed not to let NBC's actions weaken the intensity and momentum of their protest against *Beulah Land*. "We don't plan to go away," Robert Price told reporters. "The problem we have with the airing of the show will still be there in the fall." Added Saundra Sharp, "We want to see concept changes made in the script."[45]

Gerber was livid at the network's announcement. "NBC has capitulated," he charged. "A few people in a committee have brought a powerful corporation to its knees. I am embarrassed for our business because of what NBC has done; it literally has betrayed the people that work for them, broken confidence with us, made us work above and beyond normal requirement to make the air dates, and has now turned its back on us. It is the blackest day I know of in the TV medium."[46]

What did NBC expect to accomplish by postponing the air date? The network was no doubt hopeful that the protest would lose some of its momentum if the show were temporarily shelved. It is also likely that the trouble surrounding *Beulah Land* had scared advertisers away from the show. Though no one at the network would admit it, it is safe to guess that some of the sponsors of the miniseries may have begun pulling out as the confrontation escalated. Postponing it until the trouble cooled down may have been a tactic to make the commercial time on the show easier to sell.

NBC also needed time to decide how they could salvage what was beginning to look like a very expensive mistake. Since it was difficult to defend this "potboiler" on the basis of quality, and because opposition to it was coming not just from a handful of angry actors but from members of Congress and from well-respected television critics, the network decided to try justifying *Beulah Land* on the basis of historical accuracy. For this they returned to Tilden Edelstein, the Rutgers historian who had worked with writer J. P. Miller on the original script.[47]

Edelstein remembers getting a call from NBC's vice president of broadcast standards, Ralph Daniels, in March 1980 asking the professor to come down to the network headquarters in New York City. This was Edelstein's first direct contact with the network. When he arrived in Daniels's Rockefeller Plaza office, he was handed a copy of the Coalition's position paper. What the network needed, Edelstein was told, was a "rebuttal to the position paper." Edelstein was asked to read the revised script, examine the position paper, and screen the completed film. "If this is historically accurate," Daniels told him, "We're going to go with it, and we want to establish the historical accuracy." The historian complied with the request and submitted a report certifying the historical accuracy of *Beulah Land.* But that wasn't enough to satisfy the network executives. "Granted this is accurate," one of them said, "but now what are we going to do with it?" Recalls Edelstein, "They obviously wanted to get this onto the air. Their concern now was how could they cool things and protect their money interest as well as their desire to not get stuck with some sort of monster that would embarrass them. And so, we had a discussion about that and they asked me what I thought they should do."[48]

Here, Edelstein's role shifted from historical consultant to tactical advisor. He suggested that the network hire Yale historian John Blassingame. "I'd known John as a friend but I had also known him as a professional associate for a long time. He had published on slavery, and he was obviously black," Edelstein reports, "and it seemed to me that the first issue should be whether a professional historian who is black would read the thing the same way I, a white historian, read it." Blassingame agreed, and he and Edelstein began working together, evaluating the film and deciding what should be done with it. Both men shared the belief that while *Beulah Land* was not great art, it wasn't so offensive that it couldn't be salvaged. "This was not *Birth of a Nation,*" says Edelstein, "nor was it *Gone with the Wind.* We had no difficulty talking about it as something we should look at rather than something that should be scrapped." So the two began to work together to repair it.

For the most part, the historians concurred on the necessary changes, though there were a few points of disagreement. The will-reading scene which actor James McEachin had opposed during filming was one of them. Edelstein found no problem with the actor's line about not wanting to accept his freedom and leave *Beulah Land.* As Edelstein remembers, "My position was that somebody who had been a slave for forty years could have been shocked to be free and instead of jumping at freedom might say, 'Freedom, you're insulting me. I've always been free.' That was apparently too subtle an idea. John, however, believed that everybody wants to be free." Edelstein conceded on that point. The other debate was over the breastfeeding scene. Recalls Edelstein, "The question was this famous scene where Lovey (the slave girl) has a black kid and a white kid on each breast. John was saying that's an exposing of a black woman and that it is not really something that should be shown. They kept that scene essentially, but they found an outtake that moved you back a little bit, and as we jokingly said only the men in the bars now could see it on a big screen television."

All of this post-production repair work on the beleaguered miniseries took place without the direct involvement of producer Gerber or writer Miller. Edelstein believes that he played a critical role as the link between the studio and the network. "Whereas they (the network) had the right . . . to do what they wanted to since it was now theirs, they would prefer not having bloodshed and I was the one person that Gerber (and Miller) still esteemed, someone who had not sold them out." As the mediator, Edelstein would argue with Blassingame and the network to try and keep most of what Miller had written, and would periodically assure Miller that the changes being made were not substantive. Says Edelstein, "I would constantly argue to maintain what we had, and where we didn't maintain it, it was because I became convinced that Blassingame had a point, and then I would concede it. So when I conceded—'changed' is a more accurate word—something that Miller had written, I would say, 'Well, we made this concession, Jim, because I didn't think that I was right about this and it was worth dealing with.' " At one point, the network flew Edelstein to Los Angeles to meet with Gerber and "try to convince Gerber that the changes that were being made were really changes that had to do with historical accuracy . . . that we hadn't really changed *Beulah Land* in a significant way and the changes we had made were changes he could live with. . . .

[W]e were not emasculating the script or really violating his First Amendment rights. . . . It was simply a question of fine tuning."[49]

Despite these efforts, however, Gerber continued his public protest against NBC, while other members of the production community joined him in outraged chorus. On April 12, the Directors Guild voted unanimously to endorse the statement released earlier by the Caucus of Producers, Writers, and Directors, opposing the actions of pressure groups to seek prior restraint. In a dramatic gesture a few weeks later, Miller telegrammed NBC president Fred Silverman to "regretfully but urgently demand that you substitute my registered Writers Guild of America pseudonym for my name in the credits of *Beulah Land.*[50]

"If they [the Coalition] want to control what writers write," Miller told Silverman, "let them go to Russia or Iran. As it turns out they don't have to go anywhere. With your help they can do it here. When your network canceled *Beulah Land* it proved that you listened to the false allegations of the Coalition—all of which can be refuted by historical evidence—but you did not give me one minute to defend the thousands of hours of honest work done by me and the other principals in the project. You have served notice on all writers, producers, directors, and scholars that they are not safe from the depredations of special interest groups."[51]

Meanwhile, the Coalition continued to maintain its public position against the airing of *Beulah Land.* Howard Rosenberg observed in May that the group was "digging in for an extended struggle," noting that "any group that opens an office and has letterhead stationery has to be taken seriously." The move to an office in Hollywood's old Taft Building was a sign that the ad hoc Coalition was becoming a more permanent organization. Robert Price told reporters in April that the Coalition was "evolving into a Black Anti-Defamation League, which is long overdue."[52]

During the course of their protest against *Beulah Land,* Coalition leaders were learning that other advocacy groups already had long established relationships with the television industry and were often consulted about programming before it was completed. Explained Price:

> First of all, we began to learn about the standards and practices departments at the networks and the fact that they have control over the content from the beginning. We also found out that other groups had been getting courtesies that we'd never been afforded, that is, prior input. I don't know how, but we got hold of some correspondence between one of the networks and several special interest groups . . . about a particular program. . . . It was very cordial and courteous and it appeared that (the network) had flown people to New York to discuss the show. . . . I'm saying that NBC would not have this million dollar fiasco if they had related to us that way. So our position was, if that's the only excuse as to why this is not happening and this is the reason it's being called prior censorship, then we'll establish an agency that makes it not prior censorship if you contact this agency.[53]

The decision to transform the Coalition Against the Airing of *Beulah Land* into a permanent watchdog organization was also a reflection of the growing realization among the protesters that NBC was determined to broadcast *Beulah Land* and that the goal of keeping it off the air might have to be replaced by a more realistic objective. Said one of the leaders, "We had to think about what the psychological implications would be if we were not able to stop the airing; we knew that the network and the production company had invested huge amounts of money and they weren't about to lose it." At this point, two new tactics were considered. One was a boycott against the show's sponsors. But this couldn't be done until after the broadcast, since no one outside NBC knew which advertisers were planning to buy time in the program. Bonnie Allen, writing in *Essence* magazine, told her readers that, "if the coalition of organizations cannot prevent *Beulah Land* from being aired, we should watch and evaluate it, and if the film comes up wanting, scream loudly to NBC and to the advertisers who helped put the show on the air. Stay away from these products, and let the advertisers know why you switched brands. The most effective force we have is economic, and the sooner we begin to use it, the sooner we'll eliminate plantation-loving slaves from American mythology." Such optimism was not well founded, in view of other protests in the past. Seldom had any post-broadcast boycott had any discernible impact.[54]

A more promising tactic was to go directly to the affiliates and urge them not to broadcast the miniseries on their stations. The perfect opportunity came in May, when NBC held its annual affiliates meeting. This event is traditionally the time when the network presents its fall lineup with a great deal of fanfare. On May 19, while all the NBC affiliates were in Los Angeles for the meeting, the Coalition ran a full-page ad in the *Hollywood Reporter*, headlined "The Coalition Against the Airing of *Beulah Land* Welcomes the NBC-TV Affiliates to Los Angeles." The ad presented quotes from politicians and critics supporting the group's position, and invited affiliate executives to "call or visit our office" for further details about the controversy, listing thirty-six organizations who were in support of the effort. During the same week, the Coalition members demonstrated in front of the hotel where the meeting was held and handed out leaflets to seventy key affiliate stations.[55]

While on the one hand trying to convince the affiliates not to carry *Beulah Land*, the Coalition continued its efforts to be included in the network decision-making process as the miniseries was being prepared for broadcast. But over a month after NBC had announced it was postponing the airing for the purposes of "consultation," coalition leaders still had heard nothing from the network. They had expected the network to consult with them. Robert Price telephoned NBC's Jay Rodriguez in May to find out what was happening. The network executive told Price that NBC was still intending to screen the film for the Coalition. Confirming the telephone conversation, Rodriguez wrote to Price a few days later: ". . . NBC plans to show you *Beulah Land* and consult with you, after management has had the opportunity to view the program. I am not sure of the date, but we hope it will be in a month or so. I will call you in plenty of time for you to view it."[56]

Very little was heard about the status of *Beulah Land* for the next month and a half, while Blassingame, Edelstein, and the network worked on the revisions. The Coalition continued to send out mailings to the people on its list, urging them to send petitions to NBC and Columbia. On June 19th—"Juneteenth Day" in black history—Coalition members held a "town hall" meeting at the Second Baptist Church in Los Angeles to support the protest.[57]

In late June, Brandon Tartikoff, president of NBC Entertainment, met with the press to announce the schedule for the upcoming year. Asked about *Beulah Land*, the executive assured reporters that the miniseries was being carefully worked over to prepare it for broadcast in the fall. Although he tried to minimize the influence of the historical consultants, suggesting that their input resulted in only minor edits, Tartikoff maintained that the show would be substantially different—and much improved—when compared with the script which caused the original protest. "We are not putting the pages of the script on the screen," he assured the press. "There's such a thing as direction. There's such a thing as mounting a production and the way an actor interprets a role and the way the film finally gets edited, which in a lot of cases goes through a lot of changes. And there's a great deal of interpretation in reading the script." The final product would be quite acceptable to all concerned, Tartikoff assured reporters. "I think when you see it you will sit there and say 'Why all the furor?' "[58]

What Tartikoff did not explain, however, was that the complaints from the protesters were based not only on the original script. Not only had they continued to be dissatisfied with later versions supplied to them by the network, but since much of the furor over the show erupted during the filming itself, the activists had found much to be angry about in the way the show was directed. Tartikoff's remarks to the press failed to acknowledge any of this, but instead suggested that the protest over *Beulah Land* was based on material which had been revised many times over.

Since no one outside the network or production company had seen the completed film, it was anybody's guess if Tartikoff's evaluation of it was accurate. The executive indicated that NBC would be screening the miniseries to "interested parties," although who those parties would be was not made clear. "We take input from everywhere, and if they're valid concerns, we will react to them. If they're not valid concerns, we're going to stick to our guns. We are not caving in to pressure. The final shape of the program will come out of what we decide to do. Trust me on this. I wish you could see the film. It would save me a sore throat."[59]

It wasn't until nearly two months later, on August 14, that NBC announced that the final version of *Beulah Land* was ready for airing. This was the first time that John Blassingame, the black historian hired by the network, was given the opportunity to speak to the press. Interestingly, Tilden Edelstein was not included in the network press conference. Obviously, NBC felt it better to use Blassingame for these public appearances, even though Edelstein had continued to work with them on the project.

Blassingame's remarks to the press were carefully worded and less than fully supportive. The historian was careful to point out that *Beulah Land* was not representative of most plantations. "I think that what happens in *Beulah Land* is plausible," he told reporters, "but only as long as we're talking about a unique plantation, a unique set of masters. You could point to plantations that had some of the elements of *Beulah Land,* but none that had all of them." As for endorsing the show, Blassingame said, "I think as entertainment it works, but if I had to choose a film to show on slavery, this would not be the one."[60]

There continued to be confusion about the changes which had been made in the film since its postponement in the spring. The network reported that there had been very few changes in the completed production, and Gerber seemed satisfied this was true. Said Gerber: "I'm just proud that no changes have been made because of pressure from ad hoc groups and that nothing was shown to be racist." The network confirmed that the will-reading scene had been cut, but argued that any changes that were made were done so "on the basis of historical accuracy, rather than any hints of racism."[61]

Gerber later told reporters that he had made some changes himself, based upon suggestions made by historians, but he adamantly denied that any of the changes in the final product were the result of pressure. "Let me get this point over," he said, "I haven't made any changes that came out of their committee. I've repudiated their attitude from the beginning and I do right now." He did acknowledge that one small scene had been suggested by NBC's Ethel Winant, which Gerber thought was a good idea. Winant, who had been a freedom marcher in the sixties, "liked the show but wanted a third generation attitude—of young people—toward slavery," so that was included.[62]

On August 26, the network set up a screening of the completed film for the Coalition, whose leaders held out little hope they would be pleased with what they saw. The activists invited a number of people from outside their organization to attend the screening, believing that it was important to have representation not only from the Hollywood creative community but also from government, the academic community, and other political groups. About forty people attended the screening, including several academics, two representatives from the National Educational Association, and a handful of congressional aides. Historian John Blassingame was also flown out to be part of the meeting. Edelstein was not included. In addition to the blacks attending on behalf of the Coalition, NBC invited some of its own contacts from the black community.[63]

Coalition members who attended the screening held a strategy session immediately afterward at a restaurant across the street from NBC's Burbank headquarters. The consensus of the group was that, though some of the racist elements in the film had been "toned down," *Beulah Land* was still too offensive to warrant airing. They immediately drafted a statement saying the film "remains psychologically and politically dangerous."[64]

It was clear by now that NBC was not going to be deterred in its plans to air the miniseries, which was scheduled for October 7, 8, and 9. Promotional materials were prepared and distributed to the press. Rick DuBrow, TV critic for the *Los Angeles Herald Examiner*, noted that the "gorgeous brochures of lavish scenes from *Beulah Land*," which were mailed to the press, showed "no obvious scenes about slavery." Asked about this omission, a Columbia spokesperson replied these scenes were chosen because they were "the most striking" scenes in the movie. "There was no thought regarding the slavery issue," he added. "If there had been a striking picture involving that, it would have been sent out, because you can't deny history."[65]

In a last-ditch effort on September 15, the coalition sent out a "Second Position Paper Against the Airing of *Beulah Land*." Since there was little the group could do now to keep NBC from broadcasting the miniseries, the focus of the pressure shifted to local stations. And the instructions to supporters were much more detailed than the previous communiques from the Coalition. While still vowing to keep the show from airing, the Coalition recommended the following steps be taken:

1) Select a local support group, or several individuals, to work with you.
2) Familiarize yourselves with enclosed material . . .
3) Contact your NBC station manager and/or programming department by whatever means is most feasible. What you want from them: for the station to exercise its option to not air *Beulah Land* when the network airs it. Your supporting evidence includes:
   a) The racially degrading images that still remain in *Beulah Land* could stir up racial strife in your city, i.e. Miami.
   b) A Boston station set a precedent last year by refusing to air *Freedom Road* on the grounds that it was highly potential fuel for the racial crisis that city was currently experiencing.
   c) An argument against showing racially degrading images under the guise of "entertainment."
   d) An answer to prior censorship. "You don't have to wait for a gunman to shoot you, to know that a gun is dangerous."

Though continuing to label the program "unacceptable," the Coalition took credit in its new position paper for substantial changes in the revised version of *Beulah Land,* which the network had made in response to the protest:

> Congratulations! Scenes that were significantly altered because of your protest include a scene in which slaves rejected an offer of freedom in order to remain loyal to their masters. The film now shows the slaves desirous of freedom, but the offer impossible to implement under Georgia law at the time. A series of

> scenes involving a nude teenage slave girl lusting after the master's son, and her resulting pregnancy, were deleted. Many other scenes to which we had objected are still included, such as a slave woman nursing "milk brother"—one slave and one white baby—amid much glee; an adult slave reprimanding a white teenager to "start acting like you gonna be marsa" was changed from a watermelon patch to a fishing stream, with the dialog remaining intact; and a white child telling a "free" black male, "You can't tell me what to do. I'm White!" It is clear from these examples that those responsible for *"Beulah Land"* fail to understand what we mean by distorted and dangerous images.[66]

After over nine months of vehement protest against it, *Beulah Land* was broadcast on the NBC network for three consecutive nights in October. Its airing was not without complications, however. Although direct pressures had not been placed on advertisers, there were rumors that fear of controversy among advertisers had driven the cost of commercial spots in the miniseries down to "distress rates." The efforts to convince affiliates not to air the program paid off in two markets. Interestingly, WLBT, the station in Jackson, Mississippi, whose racist programming and hiring policies in the sixties had resulted in the most important court case for the media reform movement, decided that it was "not in the best interests of its community" to air *Beulah Land*. The station was now governed by an all-black Board of Directors. The station manager for WBAL in Baltimore, Maryland—another station which had faced license renewal problems in the past—told reporters that that station was not planning to air *Beulah Land* because the show was "just a fairly mediocre miniseries."[67]

The show did well in the ratings, attracting an audience of 75 million people. It did not receive good reviews, however. In fact, critics around the country spared little in their condemnation of it. The *Christian Science Monitor*'s Arthur Unger called it a "bargain basement *Gone with the Wind*" and characterized it as "simplistic and exploitative." His concluding remarks are worth noting: "The trouble with *Beulah Land* is that [it] doesn't know its place. Perhaps it was barely passable as a good old-fashioned summer print 'read' about 10 years ago. But as an electronic mass entertainment . . . liable to insult or mislead

Americans of all colors, it quite simply has no place in the television of the 1980's."[68]

Critic Tom Shales of the *Washington Post* concluded that "Beulah Land is not suitable for human habitation," adding that "we must remember to bring up idiotic, inept, cynical, exploitative travesties like this one the next time NBC President Fred Silverman is boasting about his generous contributions to the quality of life. If he had a firmer grasp on the concepts of decency and responsibility, he would go on the air and apologize for not having put this foul-minded trash down the Insinkerator where it belongs."[69]

And *Newsweek* began its review of the $10 million production with this: "Lawd forgive me, Massah Silverman, but dere's somethin' almighty peculiah goin' on at yo' plantation. Dis here *'Beulah Land'* dat NBC is showin' has a whole mess of slaves who talks jes like dis."[70]

Having failed in its efforts to keep *Beulah Land* off the air, the Coalition announced a plan to conduct a "selective buying" campaign against the show's sponsors. A few days after the broadcast, a list of those advertisers whose spots had appeared in the show was sent to everyone on the Coalition's mailing list. The campaign fizzled however, since it was too late for a boycott. The offensive miniseries had already come and gone and the protesters were exhausted from their long battle.

Few people involved with the protest believed that NBC would rerun the miniseries which had caused such an uproar. But, three years later, on September 11, 12, and 13, 1983, *Beulah Land* was rebroadcast on the network, with very little publicity. This time, black activists were caught off-guard by the show's sudden reappearance, and it was too late to organize a campaign against it.[71]

After the protest against *Beulah Land*, the Black Anti-Defamation Coalition continued to operate as a media watchdog group in Hollywood. For several years, the organization published a regular newsletter, and maintained frequent contacts with television industry decision makers. One of the group's campaigns was against ABC's comedy series *Webster*, which, like the successful NBC series *Different Strokes*, featured an orphaned black child who had been adopted by a middle-class white couple. Accusing the program of perpetuating "the image of the white person as a savior of the black child in crisis," the BAD C, as the group called itself, placed pressure on both the show's producer and the network. As a result, a special episode of *Webster* was produced to address the issue of whether or not a white child should be raised by a black family. But the Black Anti-Defamation Coalition was unable to maintain an ongoing political presence in Hollywood. Within a few years, the group had disappeared.[72]

Historian John Blassingame was retained by NBC as a consultant to advise the network on "blacks and black issues."[73]

Beulah Land was rebroadcast numerous times, not only on NBC but on cable as well. The movie was also sold into syndication for the overseas markets. If the usual practice was followed, many of the offensive scenes that were deleted for its domestic broadcasts were reinserted in the foreign version.

Cleaning Up TV

One evening in 1976, Donald Wildmon, a quiet, middle-aged minister, sat down to watch television with his family in his home in Tupelo, Mississippi. But instead of relaxing, he became increasingly enraged. The first program he tuned in contained "an explicit sex scene." As Wildmon later explained, "We don't live that way, so I asked the children to change the channel." In the next program, the characters were using "earthy" language. "We don't use that language, and teach our children not to," he said. So the Wildmons tried yet another channel. A detective story was in progress. It seemed acceptable, until suddenly, the minister recalled, "one man has another down and is working him over with a hammer." "I sat there and I recalled the changes that I'd seen in my society in the last twenty years: the magazines at the check-out stands, the movies, and other changes. And I sat there and I said to myself, 'You know, they're gonna bring this into my home.' And I became mad. I experienced some righteous indignation. . . . I became upset. And I said, 'Look, this medium has the potential to be the most constructive medium in the history of mankind, but look where it's going. . . . Television is out of whack, it's out of bounds.' "[1]

Wildmon decided to take action. He quit his post as a local preacher and set up the National Federation for Decency (NFD), adding his voice to the rising chorus against TV violence. He began publishing a newsletter, commissioned volunteers to monitor prime-time TV, and launched letter-writing campaigns to sponsors. In February 1977, NFD urged viewers all over the country to take part in a national "Turn Off TV Week." But

NFD's concerns went far beyond the violence issue, as its followers mobilized to protest the sexually permissive trends in prime-time programming. In 1979, Wildmon learned of CBS's plans to adapt the novel, *Flesh and Blood* into a TV movie. The story involved a young man's incestuous relationship with his mother. NFD orchestrated a protest against the film, picketing CBS headquarters and writing to the top 250 advertisers in TV, warning them about the consequences of sponsoring the movie. The network postponed the airing, blaming production problems, but finally broadcast the film in October, with the most explicit scene edited out.[2]

Though it had succeeded in stirring up some trouble over the film, Wildmon's watchdog group did not appear to have a broad enough constituency to pose a serious threat to the networks. A *TV Guide* article characterized the National Federation for Decency as a "one man crusade to 'clean up' TV," with a tiny band of followers, and described Wildmon as "founder, chief executive, corresponding secretary, top publicist, and No. 1 fundraiser, missionary, speechmaker and pamphleteer. . . ."[3]

Wildmon was considered a nuisance in 1979. But by the following year, the minister from Tupelo was taken much more seriously.

Conservative groups needed the right moment to launch an effective public campaign, a time when the press would take their efforts seriously, when they could garner sizable grassroots support, and when they could be perceived as a sufficient threat to the powerful television industry. That moment came in the fall of 1980, when the national election resulted in a landslide victory for conservatives, indicating that American public opinion was shifting further right. Conservative organizations were beginning to play a bigger role in American politics. One of the most powerful of these groups to rise to prominence during that election year was the Moral Majority. A coalition of religious and political groups led by TV evangelist Jerry Falwell, the Moral Majority used the technological mass mailing wizardry of Richard Viguerie to mobilize broadbased support and generate large sums of money for its cause. With this effort, the organization supported conservative politicians, who were able to win several important congressional seats from liberal incumbents.

Conservative leaders knew that the media issue was one with the power to galvanize their constituency. For years, right-wing critics had charged the media with left-leaning bias. Groups like Accuracy in Media had repeatedly criticized network news organizations for their alleged liberal bent. Vice President Spiro Agnew had garnered much support among conservatives in the late sixties with his public accusation that the news media were controlled by an Eastern, liberal establishment. To conservatives, recent trends suggested that entertainment television was run by loose-moraled liberals who insisted on forcing sexually explicit, anti-Christian material into the homes of middle Americans, threatening their cherished value system.[4]

Attacking television also had the potential of attracting broader support. Only a few years before, the issue of TV violence had successfully mobilized masses of Americans to take action against television. Right-wing leaders no doubt sensed that part of the momentum of the anti-violence movement stemmed from a much deeper public alienation from television.

Once the public tide appeared to have shifted, the Moral Majority did not hesitate to take action. In a move reminiscent of the PTA's earlier call to battle, Moral Majority leaders issued a stunning announcement on the day of the election. They had committed more than $500,000 to launch a "nationwide campaign against excessive TV sex and violence." Proclaiming that television should "reflect the morals of this country," the leaders warned the networks that they'd better start programming "to all America, not just the homosexuals and the free and extramarital sex crowd." Using sophisticated, computerized mass-mailing techniques, the Moral Majority claimed its mailings reached 480,000 people, including 72,000 members of the clergy.[5]

A few months later, a larger, more ominous coalition of activist groups emerged. In February 1981, the National Federation for Decency combined forces with the Moral Majority to form the Coalition for Better Television, with Donald Wildmon at its helm. Joining the new coalition were some of the most powerful organizations of the so-called "New Right," including Phyllis Schlafly's anti-ERA group, Eagle Forum; the anti-abortion organization, American Life Lobby; Christian Voice; and the Clean-Up-TV campaign headed by Rev. John Hurt. Each of these groups had waged individual campaigns against network television, but this was the first time a large coalition of organizations on the right had banded together with a clear strategy to call for major changes in program content. In his new leadership role, Wildmon now claimed a constituency of major proportions. "We could conceivably be talking about 5 million families," the minister told the press, poised to "move like a gigantic tidal wave."[6]

Coalition leaders characterized themselves as reformers acting on behalf of the public interest. "We seek to promote constructive television," Wildmon explained to the press. "By that we mean programming that adds to man's cultural, social, mental, emotional and spiritual heritage in an informative and entertaining manner, and which does not appeal to nor exploit man's prurient nature . . . that which lifts and inspires, not that which degrades and exploits. . . . We're not against sex or against violence on television. We're against the excessive and gratuitous."[7]

But the cause was also framed in fervently religious terms. This was no less than a holy war, CBTV leaders proclaimed. "The ugliest mark," said Wildmon, "that will be written against the institutional church in this century will be its silence in the face of immorality and the decline of public morals. Our society promotes hedonism, heathenism, materialism, greed—everything that's destructively different from Christian values. Yet the Church is silent."[8]

The campaign relied heavily on the press to dramatize the threat of this new move against television. At strategic moments, CBTV leaders called press conferences, issuing dramatic announcements about the coalition's next planned attack. The issues were inherently newsworthy; the New Right was an emerging political movement, and controversy over television had always generated public attention. Consequently, the press dutifully covered the Coalition's every move. Newspapers quoted Wildmon frequently. As Todd Gitlin observes, "For months, reporters uncritically relayed Wildmon's claim that the coalition included 200 groups—later boosted to 300, then to 401. Many were local letterhead organizations, but it took until June before a CBS correspondent reported that, of a sample of some sixty organizations and individuals whom Wildmon counted in CBTV, about 30 percent denied membership." In an ironic but predictable twist, television instantly transformed Wildmon into an overnight media celebrity by featuring the crusading minister on talk shows. The coalition also took advantage of its own religious TV networks to advance the cause and get the word out to its followers.[9]

Because the coalition called for sweeping changes in primetime content, its strategy circumvented the network system for managing advocacy groups. Instead, coalition leaders borrowed the pressure model used by the anti-violence forces, adapting it to fit their own purposes. Having learned—from their own experiences as well as observation of other groups—which tactics worked and which would not, CBTV chose not to play by the established rules of the game. "As far as letters are concerned," Wildmon told his followers, "forget writing to the networks. As long as the money is coming in, they don't care what you think. . . .The real clout that churches have is against people who pay the bills—the advertisers, the sponsors." To fight this holy crusade, the coalition was assembling a sophisticated computerized operation. A monitoring program was set up, using an army of 4,000 volunteers from across the country to count "lewd remarks, sexual innuendo, outrageous acts," and other instances of "gratuitous sex."[10]

But while media coverage was essential to CBTV's success, secrecy and suspense were also key features. When asked to explain the details of the monitoring system, Wildmon refused. Better to keep the process a secret than to expose it to criticism. The computerized monitoring results would then be used to identify sponsors. But unlike the anti-violence groups—which regularly released lists of sponsors identified with violent programming, CBTV kept the industry in suspense. The one sponsor found most guilty of supporting immoral and offensive programs would not be identified until June 1981. At that time, CBTV would immediately launch a nationwide boycott against the targeted company. On PBS's *Inside Story,* Moral Majority leader Jerry Falwell vividly described the planned scenario:

> We have made a commitment . . . to put two million dollars at least into the boycott. For example, if we [the Moral Majority] were called and told, "Okay, the boycott begins X date," we would write to four and a half million families, instantly, saying, "Here are the facts, here are the people, here are the products, here are the things we don't want to buy anymore, here's why," and so forth. So would all the other three hundred organizations. We would buy full-page ads across the nation, listing the Public Enemy Number One and his products, and urging the boycott. And we would get on TV and radio and press conferences everywhere and eighty thousand pastors next Sunday would preach on it, and would put it in the Church bulletins and within thirty days the dust would be flying.[11]

NBC's network president Robert Mulholland tried to play down this threat by lumping the Coalition with all the other "special interest groups" his network regularly dealt with. In a March 1981 press interview, Mulholland listed a number of other religious groups, and cited

> another preacher—the Reverend Cecil Todd of Revival Fire Ministries in Joplin, Mo. Todd hit the headlines in 1975 when he hand delivered 750,000 letters protesting immorality on the tube. He rolled into New York in a van and with news cameras blazing, dumped sacks of 250,000 each at NBC, CBS, and ABC. Executives of all three networks met with Todd and listened to his complaints, but nothing changed in the fall lineup.[12]

But despite industry assertions to the contrary, there was cause for genuine alarm. Though business had pretty much returned to normal after the anti-violence battles a few years earlier, the memory of the intense pressures and the disruptions they had caused was still strong.

There were significant differences between the anti-violence issue and the issues of the New Right, but they did not make the new campaign any less disturbing to the industry. The scientific community and Congress had made the anti-violence groups an imposing threat to a regulated industry. Perceived by many as a serious public health issue, TV violence had brought together a wide range of politically diverse groups and appeared to draw broad grassroots support. Though the Coalition for Better Television lacked the scientific and governmental legitimacy of the anti-violence forces, it had the strong political backing of a newly fortified conservative movement. Conservatives had demonstrated in the recent election that they were a political force with sophistication and the capability to mobilize large numbers of people. Unlike the organizations that had organized around the TV violence issue, the coalition was focused not on one kind of potentially harmful content but rather on an entire value system. Though CBTV leaders couched their concerns in similar terms to those used by anti-violence groups, many feared that the real agenda of this right-wing group was to suppress ideas.[13]

The networks waged a counter-attack in the press. CBS senior vice president Gene Mater was one of the more outspoken critics of the coalition. Calling the coalition both "unscrupulous" and "formidable," Mater told an industry gathering: "[Their] idea is not to attract sympathy and support [for their cause] but to compel concessions. This involves the violation of rights, not the exercise of them. The advertiser is the pressure point. The concession comes down to making TV do and say and show what [the coalition] thinks TV should do and say and show." "We have no problems with a teacher teaching or a preacher preaching not to watch a particular program," the executive told the press, "but we have problems with a threatened boycott. . . . No matter how you couch it, and no matter how well intentioned it is, that's censorship."[14]

Boycotts were nothing new as a political tactic. They had been used successfully throughout the civil rights movement against those companies that refused to comply with desegregation laws. Farm workers' unions had waged boycotts of produce to force farmers to improve working conditions. Boycotts of television sponsors were also a longstanding practice, though until recently they had been infrequent and ineffective. But never before had any group or coalition of groups posed as big a boycott threat as the Coalition for Better Television. The right-wing groups in the fifties had used the threat of boycott as an effective pressure tactic. But, though many advertisers cooperated with them, these groups lacked the sophisticated machinery to engineer a nationwide boycott. The boycotts over *Amos 'n' Andy* and *The Untouchables* may have been partially successful, but in both cases, the number of participants was relatively small. Though the PTA—with its grassroots constituency and high visibility—did have the capability for an effective national boycott, its leaders chose not to single out one company but rather to release lists of those sponsors identified with violent program content. The energy and momentum behind the campaign was therefore diffused. While participants in the campaign were free not to buy the products of these companies, no coordinated boycott was ever launched.

The Coalition for Better Television, on the other hand, had both the capability and the determination to wage economic war on advertisers. The decision to focus all boycott effort on one company made the threat even greater. If, indeed, CBTV had the followers it claimed to have, real financial loss was a serious possibility. "If only one million families participated in a boycott directed against a manufacturer of toothpaste," one industry executive later explained, "it could have a significant effect. And if the boycott were extended to the range of personal care products, the cumulative effect could reduce a company's sales 5 to 10 percent. Not enough to bankrupt anyone, but enough to make management unhappy."[15]

The coalition's pressure campaign sparked a public debate during the next few months about the legality and ethics of sponsor boycotts. Wildmon repeatedly defended his group's actions, labeling the boycott plan as a form of "militant consumerism." "The network has the right to spend its money where it wants to; the advertiser has the right to spend its money where it wants to; and the consumer has the right to spend his money where he wants to. The clearest expression of the first amendment is the right of a corporation or individual to spend money where they want to." Participating in the boycott was a moral duty, the minister told his followers. "It's the responsibility of every Christian to be a good steward of all his money," Wildmon instructed. "Censorship has existed ever since man learned to communicate. It has to exist of necessity. It exists right now. Everything you see on television has been censored; but the question is, who is doing the censoring? When the networks cry censorship, they're usually speaking of an official act. We're not official, not a government organization. We're a citizens' group. The only clout we have is with personal economics, and that's a perfectly legitimate clout."[16]

This argument was echoed by conservative columnist George F. Will, who suggested that boycotts were the only effective means for conservatives to be taken seriously by the networks. "It is hard (and hardly obligatory)," he wrote, "to credit sincerity of people who shout 'Censorship'—with that word's connotations of coercive state action—when people are simply planning to practice selective buying of beer and pantyhose. A network spokesman decreed that boycotts are 'censorship' and are 'a totally unacceptable method of trying to influence programming.' I called to ask what he considered acceptable. He said—are you ready for this? 'Oh, writing a letter.' "[17]

Law professor Alan Dershowitz countered this position. Though not ruling out sponsor boycotts altogether, he argued that, "although most boycotts are constitutionally protected, some of them are morally wrong." At an industry gathering, he suggested some criteria for determining the appropriateness of sponsor boycotts:

> Surely it is more appropriate to boycott an advertiser who plays an active role in determining content than one who plays no role. If, for example, a sponsor were to say, "I'll advertise on this show only if there are more exposed breasts, or it puts down gays, or if it casts a Klan member as Martin Luther King," then the propriety of an economic boycott becomes more obvious. But if the sponsor merely declines to remove his ad from an objectionable show, the propriety of a general product boycott becomes more questionable.[18]

While the public debate continued over the ethics of the coalition's tactics, network executives planned their own counterattack designed to undermine the potential impact of a boycott. As the PTA had done in 1977, coalition leaders had launched their public campaign during the critical period when advertisers make the "up-front buys" in the next year's prime-time schedule. Since the pressure campaign was bound to affect advertiser decisions, the networks focused their efforts on assuring the advertising community that they had very little to fear. Both ABC and NBC commissioned special audience surveys to determine the extent of the boycott threat.[19]

ABC distributed the results of its study (designed by its social research department and conducted independently by the National Survey Research Group) via closed-circuit TV to 1500 advertising representatives. When the network presented its prime-time schedule in May, it also summarized the research results for worried advertising executives. "More than 91% of Americans," the network officials assured the advertisers, have never at any time in their lives boycotted a product because of something in television they disliked. Slightly half of the respondents in the ABC study had heard of the Moral Majority, while 6.6% of them reported themselves as members. Of those who said they were members, 44% indicated they supported "the Moral Majority's attempts to influence programs to conform to their standards and values." Twenty-nine percent opposed this tactic and 27 percent were undecided. The survey also showed that those people claiming to be members of the Moral Majority watched almost as much "immoral" TV as nonmembers. "Among those calling themselves Moral Majority members," the *New York Times* reported, "13 percent say they were regular viewers of 'Soap' versus 13.8 percent of the total population. For 'Dallas' the figures were 33.7 percent for Moral Majority members and 35.8 percent of that total population."[20]

A more disturbing finding was that a sizable number of Americans in the general population—20.5 percent—reported they would support a group seeking to influence television programming by using tactics that included boycotts. But the network chose not to include this information in its report that was released publicly the following month. When asked later about this, a network official argued that this did not mean people actually would do what they said they would do.[21]

NBC employed the Roper Organization to conduct its study to find out how the general public felt about some of the programs the coalition opposed. Though many people expressed concern over excessive TV sex and violence, very few thought the specific shows which the coalition had singled out exhibited those characteristics. "The majority of the viewing public doesn't share the criticism expressed by the leaders of the coalition," the report concluded. "There is little dissatisfaction with the treatment of sex in these programs, less dissatisfaction with violence, and even less sentiment for taking these programs off the air because of 'sex, profanity, or violence..' " But advertisers were still not convinced. Nor were many others in the industry.[22]

In May the Television Academy sponsored a symposium entitled "The Proliferation of Pressure Groups in Prime Time" in Ojai, California, a quaint resort town near L.A. The meeting of about sixty producers, writers, network executives, advertisers, and scholars had been planned before the Coalition for Better Television launched its attack, and it was supposed to address the range of special interest groups that had been attempting to influence network programming. But since the gathering was held less than sixty days before coalition's target date for a boycott, that group dominated the three days of discussions. "Although a list of 132 special interest groups," wrote Howard Rosenberg of the *Los Angeles Times*, "—ranging from the PTA to the American Psoriasis Society and the American Egg Board—was distributed at the symposium, the Wildmon/Falwell group cast a shadow over the entire affair." Other groups were mentioned, but none received the attention, or generated the heated debate, that the coalition did.[23]

It was clear throughout the meeting that, despite what the industry was saying publicly, advertisers were intimidated by the coalition. In private conversations, advertising executives speculated about which company was likely to be the group's target. "We are nervous," said F. Kent Mitchell of General Foods. "At a time when the public is very uncomfortable about losing control of their lives and with something called the new Moral Majority coming in, [General Foods wants to] alienate as few people as possible." Added another: "I don't think my company's going to be nailed, but God help the companies that are." And many advertisers seemed to feel that the networks were not in an entirely defensible position. "There's been an 'escalation of sex' in prime time," one executive noted. Added another: "We don't feel the industry comes to this debate with clean hands, because of gratuitous sex." Observed Robert Goldstein of Procter and Gamble: "There are an awful lot of Americans who feel that even if they don't totally agree with Donald Wildmon, he has a point," And Grant Tinker, who was soon to become president of NBC, acknowledged that, "if we cleaned up our own act, we wouldn't have others trying to do it for us."[24]

And indeed advertisers had begun to translate some of their fears into action. Using the same mechanisms that had been put in place to protect them from pressure over violence, advertisers were able to flag content that could make them vulnerable for boycott from the coalition. A number of them were pulling out of questionable programs. "There's been an acceleration in the last several weeks of sponsors saying they want to stay off certain [ABC] programs," ABC's Alfred Schneider admitted. CBS's Gene Mater conceded that his network, too, was experiencing sponsor withdrawals.[25]

Though no one knew yet who the boycott target would be, the coalition offered to let advertisers know how they were doing. All they had to do was call the Tupelo headquarters, where a computer printed biweekly status reports of advertisers scores. And some advertisers were keeping in touch. Dow Chemical, reported the *New York Times,* confirmed that it had been checking with the coalition, but "would not say if it had been influenced."[26]

There had been speculation that some companies might be directly negotiating with the coalition. But the real bombshell came on June 15, when Procter & Gamble, one of the largest sponsors in network television, announced it had withdrawn sponsorship from fifty network television shows during that season because they contained excessive sex, violence, and profanity. "We think the coalition is expressing some very important and broadly held views," P&G's chairman Owen Butler told an industry meeting. "I can assure you we are listening very carefully to what they say and I urge you to do the same."[27]

Though many were surprised that P&G executives went public with their decision, they had good reason to clear their name with the coalition. One of the more conservative companies in the business (and one of the first to have program content guidelines), Procter & Gamble had raised the ire of right-wing groups with its 1977 decision to bail out the beleaguered NBC movie *Jesus of Nazareth.* Since then, Wildmon's National Federation of Decency had targeted the company as a top sponsor of sex and profanity. In an open letter to then P&G chairman Edward G. Harness, Wildmon wrote in 1980, "We will keep Procter & Gamble in our prayers, and off our shopping lists with hopes that your good name will not continue to be associated with television immorality." None of this was mentioned in Wildmon's statement to the press after Procter & Gamble's announcement. "They were not a company we were considering boycotting, based on our monitoring," Wildmon said. "I think Mr. Butler was reflecting the private sentiments of nearly every major advertiser, because we've talked to them. I think this is just evidence we're being heard, and that our concerns are shared by other people."[28]

Procter & Gamble's move marked a critical point in the confrontation between the television industry and the coalition. If there had been doubts before, it was now evident that advertisers were willing to give in to pressure. And they were doing more than just withdrawing some of their ads from the most controversial programs. As they had done during the anti-violence campaigns, advertisers were putting their own pressure upon the TV executives who controlled the content of prime-time programming. "Over the past six months, as the coalition has really come to the forefront," one ad executive explained, "I think more advertisers have been talking directly to the networks, and through their agencies about the quality and content of programs." As more and more advertisers were becoming increasingly selective about which programs they would sponsor, there was an added economic pressure. Because of supply and demand, the cost of a thirty-second spot in so-called "clean" shows became more and more expensive because there were so few shows remaining which were acceptable.[29]

This capitulation by sponsors was alarming to many people, both inside and outside the industry. The Moral Majority's ascendance to political power had already spawned a new counter-group, People for the American Way. Its founder and leader was producer Norman Lear, and its board included a number of prominent religious leaders. People for the American Way announced a counter-attack against the Coalition for Better Television. Launching its own national campaign against "the intolerant messages and anti-democratic actions of Moral Majoritarians," the group planned to send public service announcements to all the television stations around the country to oppose the Moral Majority.[30]

But these efforts could do little to reverse the impact the coalition's pressure had already had on advertisers. By the latter part of June—with the planned boycott only a few days away and the target sponsor yet unnamed—coalition leaders held a series of secret meetings with advertising executives at an undisclosed location in Memphis, Tennessee. Why had the advertisers agreed to the meeting? "One reason . . . was to see what they wanted," a spokesman for Warner Lambert said. "It's been our longstanding posture to be willing to sit down and talk with responsible groups and parties interested in the activities of the corporation." But obviously the advertising industry hoped to avert a boycott.[31]

In a Moral Majority mailing on June 26, Jerry Falwell told his followers the group was "one step away from victory." "A majority of national advertisers have openly expressed a desire to cooperate with the Coalition for Better Television in cleaning up television and making it fit for human consumption," he wrote. "Only a few companies are refusing to clean up their act." But "the networks are adamantly resisting all pressures from the American people and are acting very arrogantly. . . . So, we cannot pull back now." Followers were instructed that a press conference would be held in the next few days to announce "a national boycott of the company (or companies) violating the American traditional value system." The letter ended with an urgent plea for money to support the campaign. "Without your help, this boycott campaign will never get off the ground, and the pornographers will capture the airwaves forever. . . . Please sit down right now and write the largest check you can to Moral Majority, Inc. Remember, in return for your gift, I'll send you a transcript of next week's press conference for your personal file."[32]

But the following day, a spokesman for the Moral Majority told the press that the coalition was "leaning toward" postponing a boycott because of successes it had achieved in its recent meetings with advertisers. "We've been able to achieve nearly everything we set out to achieve without a boycott," explained Cal Thomas, communications director for the Moral Majority. "Peace is always better than war, and discussion is always better than confrontation. The networks and the advertisers seem to have recognized that they have a moral and ethical responsibility to the public." Thomas further explained that, in the meetings the group had held with advertisers, most of whom were unidentified, "all of the companies we met with indicated that they would reevaluate their advertising policies with a view toward rewriting them. 'Only one company that the coalition contacted so far has failed to respond.' " The Procter & Gamble statement of ten days before was credited by the coalition's leaders as a precipitating event opening up the negotiations. "It's like the prime rate," said Thomas, "when No. 1 announces, a lot of others follow. The response we've gotten from advertisers since the Procter & Gamble speech has been very gratifying."[33]

Some observers suggested other reasons for the postponement of the boycott. In addition to the studies done by the networks, several independent surveys indicated little support existed for the planned boycott. One ad agency executive assessed that "they suddenly discovered that a boycott was not worth a damn thing. It was a tactical move so [the coalition] would not face the embarrassment of failure."[34]

Since the pressure campaign had been most intense during the period when the fall schedule was being finalized and most prime-time shows were still in production, it was bound to have an impact on program content. And it did. As Todd Gitlin observed at the time: "The fall schedule does not include a new crop of jiggle shows (and *Charlie's Angels*, the mother of all jigglers, is gone)." Instead, one advertising executive told Gitlin, the pendulum seemed to have swung back in the direction of "more cops and robbers." A less perceptible effect of the campaign may be its impact on borderline shows, Gitlin suggested. "What happens at renewal time to the marginal show which has been plagued by a lot of last minute advertising pullouts, especially if the sales department reminds programmers that the spots had to be sold at bargain rates? What happens to the risky pilot, the one that includes some language the Reverend

Wildmon may think is profane, or some action his monitors may think amounts to 'implied sexual intercourse,' or a social situation that CBTV may regard as intrinsically risque?"[35]

The pressure campaign had a particularly strong impact on a new ABC comedy series that premiered that fall. *Love, Sidney* starred Tony Randall as a middle-aged homosexual who takes up residence with an unwed mother. When right-wing leaders heard of the network's plans for the new show, they alerted followers and singled out the new series for attack. A *TV Guide* article about the gay lobbyists in television confirmed the minister's suspicions about the control of prime-time programming. Shocked to learn that "the networks have been submitting scripts involving homosexuals to a gay organization for reviewing and editing," Falwell concluded that, "obviously, the networks are more concerned about offending homosexuals than moral Americans." Due to the special pressures directed on this show, by the time the new series hit the air, the homosexuality of the main character was almost impossible to identify.[36]

The coalition continued its vigilant pressure campaign into the 1981–82 season, reminding the industry that the threat of a sponsor boycott was still very much there. And the press continued to pay attention to Wildmon. "If I were a betting man," Wildmon told a New York *Daily News* reporter, "I'd bet that before the 1981–82 television season is over, there's a 90% chance there'll be a boycott." This time the coalition had broadened the scope of its concerns to include "sex-oriented" ads. Not only was CBTV holding the advertisers responsible for the programming they sponsored but it was also planning to scrutinize the advertising message themselves. The advertising industry was by no means off the hook.

Though the June boycott had been averted, advertisers remained on their guard. A public relations executive for advertisers advised the industry to brace for a boycott and offered steps that could be taken in the event of a boycott:

> At the very least, companies should review the programs on which their advertising appears as well as reviewing their advertising content and objectives. Such a review should not be done for the sole purpose of placating any external pressure groups, but rather with the view that advertising and program

> content meet explicitly defined and defendable goals. Second, companies should assess their vulnerability to boycotts. While IBM and Xerox are relatively invulnerable, some kinds of consumer products companies are very vulnerable indeed. Third, companies should set a strategy to deal with or avoid a boycott. When the list is published and the boycott begins, panic too often enters.[37]

If advertisers needed an added reminder of the coalition's presence, Wildmon provided it by mailing letters to them warning of the inevitability of a pending boycott, and enclosing sample magazine and newspaper ads it planned to use to promote the boycott of specific products. The letters ended with an offer from CBTV to "work with any advertiser toward correcting the situation and avoiding confrontation."[38]

By January Wildmon had stepped up his threats, setting a March 2 date for the boycott. But there were two notable changes. First of all, the coalition had shifted its focus. Borrowing the rhetoric of minority groups, CBTV was now demanding fairer treatment of Christians in prime-time programming. Accusing the networks of "secular supremacy" and "Anti-Christian bigotry," Wildmon charged that "Christian values are censored out" of entertainment television. But more important than this change in emphasis was the fact that the coalition had lost its most important members. The Moral Majority had chosen not to participate in this new boycott."The networks have made a good-faith effort," spokesman Cal Thomas explained, "and our belief is that it's just not appropriate to turn around and start a boycott." ERA opponent Phyllis Schlafly had also taken her group out of the coalition's boycott plans.[39]

With the Moral Majority missing from its ranks, the diminished Coalition for Better Television was no longer the fearsome force it had been the year before. What was noticeably different about the new boycott threat was the dramatic absence of hysteria in the industry. Although advertisers had prepared themselves for a boycott, absent were the speeches and surveys from the year before. Instead, the industry exhibited just a quiet wait-and-see attitude.

Wildmon's next two major moves further reduced the coalition's power to pressure the television industry. As long as the boycott had remained a threat and the target company had been kept secret, the group had been able to intimidate advertisers effectively. But as soon as the boycott was called, the coalition risked having it fail. The plans were further hampered by the choice of a target.

Wildmon announced on March 4 that CBTV was boycotting not an advertiser, but the NBC network and the company that owned it—RCA. The reason? Because, Wildmon explained, NBC's shows were worse than those of the other two networks. Some people speculated that CBTV had selected RCA/NBC because it was in financial trouble. Ratings on the network had been down considerably in recent months and RCA had been experiencing some financial difficulty. One other possible explanation for the new boycott target was that the coalition, in its "dialogues" with the advertising industry which had started the previous spring, had worked out a deal.[40]

CBTV quickly learned that a boycott against a major conglomerate like RCA could be a complicated venture. The day after announcing which products its members should avoid buying, the coalition had to issue an apology to Gibson Greeting Cards which had mistakenly been included in the list. What the group had not known was that the greeting card company had been sold by RCA to another conglomerate four months earlier.[41]

It would take a while to see what impact, if any, the boycott would have on RCA sales and NBC ratings. In the meantime, Wildmon's next target was one of NBC's TV movies. *Sister, Sister* was a story of the deep emotional conflicts among three black sisters in the contemporary South. One of the major characters was a black minister who steals and has affairs with two of the sisters. Wildmon accused the show of being "anti-Christian." The irony of this protest was that this particular movie had been on NBC's shelf for three years and was scheduled only after some prodding by black groups. The circumstances by which Wildmon had found out about the movie were also ironic.[42]

Wildmon had been invited in May to a seminar on "Special Interest Groups and the Media," sponsored by the Aspen Institute for Humanistic Studies. The meeting, held high in the mountains in Aspen, Colorado, was a kind of follow-up to the earlier Ojai meeting. But while no groups had been invited to the first meeting, advocacy group leaders were the special guests at this all-expenses-paid, three-day gathering, which also included representatives from the three networks and a handful of people from the creative community. The purpose of the seminar was to work out some "rules of the game" for advocacy groups to follow in dealing with film and television. Several films were screened for discussion. One of them was *Sister, Sister.* Though Wildmon participated in the meeting, he was not present during the screening. He did hear about it, however, in the next morning's discussion session. Unhappy with what he heard about the film, Wildmon used his tried and true tactics to launch a protest. He fired off letters to hundreds of advertisers, accusing NBC of "Anti-Christian, anti-religious stereotyping" and urging potential sponsors to refrain from participation in the upcoming broadcast of *Sister, Sister.*[43]

This time, the network counter-attacked. Standards and practices executives enlisted the help of black activist groups—who strongly supported the movie—to urge the advertisers who had bought time not to pull out. Though some of sponsors did withdraw, the effort convinced others to stay with the broadcast.[44]

Wildmon clearly had chosen an inappropriate target. *Sister, Sister* was neither a typical network exploitation piece nor did it contain excessive sex or violence. It had been written by noted black author Maya Angelou, and was considered a serious piece of TV drama. When Angelou was told that Wildmon, who never had seen the movie, had publicly described it as "anti-Christian," her response to reporters was, "Now that is ignorant. This silly man, indeed."[45]

Though the protest over *Sister, Sister* may have cost the network some money, it took a much greater toll on the Coalition for Better Television's credibility. As one viewer wrote to the *Los Angeles Times:* "I'm a little tired of these 'so-called Christians' who would deny freedom of expression and freedom of choice. As a Christian woman and mother, I have always preferred that my children view wholesome programs with happy endings where good overcomes evil. However, I realize that life is not all good and all bad; that some people experience problems and life situations that I would never imagine, and I feel their side should be depicted also. How can people grow and become multi-dimensional, caring, informed, empathetic persons if only one aspect of life is allowed for viewing?" Wrote another viewer: "*Sister, Sister* is superb television, pointing the way to what television could be were it not for the spinelessness of the major advertisers who control the medium when confronted with pipsqueaks of the Wildmon variety."[46]

And what of the boycott? Had any impact been felt by RCA or NBC? Wildmon claimed in November 1982 that the boycott had already hurt the company, taking responsibility for a large decline in third quarter profits for RCA's consumer electronics division. Not so, countered an industry spokesman. "The reason those earnings are down is because of excess [foreign] competition, not the boycott." Overall, RCA profits had improved, it was reported in January 1983. The profit for the full year 1982 was up 75 percent. Moreover, the network earnings had improved dramatically since the start of the boycott, ending a four-year slump. By May, the *Los Angeles Times* reported that "NBC turned the corner in 1982, more than doubling its profits from $48.1 million to $107.9 million. Judging from the first quarter returns this year, it looks as though profits will double again in 1983."[47]

CHAPTER NINE

The Hollywood Lobbyists

The guests had finished their catered buffet dinner of "nouvelle Chinese" cuisine, and had found their seats in the large, comfortable living room of this Spanish-style home in the exclusive Hancock Park district of Los Angeles. The atmosphere was friendly and informal, and most of the guests already knew each other. This "gathering" of about forty prominent Hollywood producers, writers, and directors was sponsored by a nonprofit organization called Microsecond. Microsecond's director, Norman Fleishman, a soft-spoken man in his late forties, introduced the evening's guest speaker, Michael Pertschuk, public interest advocate and former chairman of the Federal Trade Commission. Pertschuk spoke for about twenty minutes, telling stories of successful public interest lobbying campaigns. Like the standard "pitch" that producers make to programming executives, Pertschuk's talk was spirited and entertaining, packed with drama, humor, and colorful characters.

The audience included some of the top people in television's creative community. Interested in political and social issues, they wanted a chance to meet and talk with someone who had been on the front lines of political action. Always looking for new material, they listened carefully, quickly processing what they heard. Was there something here that could be worked into one of their current TV series? An idea for a Movie of the Week?

This 1985 meeting was not the first of these events. Norman Fleishman had been holding these soirées in Hollywood since 1973, when, as the director of the Los Angeles chapter of Planned Parenthood, he had given his first party in support of producer

Norman Lear during the *Maude* controversy. During the pivotal 1972–73 television season, when entertainment television's new trend toward more liberal social and political issues was under fire from conservative political groups, Fleishman was on hand in Hollywood to orchestrate support for those who took the risk of incorporating such issues in their TV shows. Since that time, Fleishman had become an institution in Hollywood. Some called him a "liberal pied piper." After leaving Planned Parenthood, Fleishman became TV project director of the Population Institute. Throughout the seventies he continued to hold regular gatherings in the homes of Hollywood celebrities. Noted speakers such as Norman Cousins, Paul Erlich, and Margaret Mead were featured. In recent years, Fleishman had set up his own nonprofit organization, Microsecond, to serve as "a catalyst in mass media entertainment programming to wake up the public to the threat of nuclear annihilation."[1]

By the mid-1980s, Norman Fleishman had been joined by a handful of "Hollywood lobbyists" who sought to influence entertainment programming by going directly to producers, writers, and directors. Though their styles varied, these groups shared a common strategy, one that was strikingly similar to that used by government lobbyists to influence Congress. These nonprofit organizations situated themselves as close as possible to the centers of decision making, setting up offices in Hollywood where they could establish and maintain routine contact with the creative community. Like Washington lobbyists, they made themselves valuable by providing facts, expertise, and support. Because their success depended on Hollywood's willingness to work with them, their tactics were cooperative rather than confrontational. They successfully adopted the rules of the game, compromising when necessary, and adapting to changing industry trends. Some of them offered incentives in the form of awards and public recognition.

Like the organizations that had successfully negotiated ongoing relationships with network standards and practices departments, the Hollywood-based groups were advocates for "manageable" issues. They did not demand broad changes in programming content, but sought narrowly defined alterations in the patterns of prime time. Unlike most other advocacy groups, however, these organizations focused their efforts primarily on the introduction of new ideas to television. Since most of them represented those social causes that could be easily incorporated into entertainment programs, they sought ways to participate in the generation of programming. This approach required a familiarity with not only the structure and process of decision making in network television, but also the unique imperatives that govern the development of programs. It involved knowing what "works" in entertainment TV and how issues could be translated into dramatic elements. It meant learning how to frame issues so that they conformed to standard formats, commercial constraints, and audience expectations. It also required a full understanding of the culture of Hollywood.

The production of prime-time programming is controlled by only a few companies. They operate within a tightly knit society, where interlocking social relationships are an integral part of the structure and operation of business. The community—as those in the industry call it—is not easily accessible to outsiders. For advocacy groups, it is particularly difficult to penetrate, since producers and writers have traditionally resisted efforts by outside organizations to tell them what to do. In order to gain access to the creators of entertainment TV, Hollywood lobbyists attempted to find their way into the informal social structures of the production community. They did this by initiating contacts with a few producers who were sympathetic to their issues and then building upon those relationships. Though some of these groups also worked with the network standards and practices departments, all of them based their strategies for influence on direct involvement with the production community. A few of them formalized their relationships by setting up advisory boards comprised in part of industry representatives and by linking up with industry guilds and organizations.[2]

Though some of the groups had only brief periods of involvement with Hollywood, others became permanent fixtures, solidly integrated into the world of TV production. With continued funding they were able to remain a significant presence. The glue of permanence was a compatibility—of style as well as content—between the advocacy groups and the Hollywood creative community. Smooth relations between individual producers and writers and advocacy group representatives were enhanced by a congruence of values and political attitudes. The more successful Hollywood lobbyists were also able to "fit in" to the Hollywood culture, successfully internalizing the "mind set," the routines, and the rituals. Though they remained outsiders, they learned to negotiate the meetings, lunches, and social events so essential to the functioning of business in Hollywood.

Population Institute/Center for Population Options

Advocates for population control pioneered the strategy of appealing directly to program creators. The New York-based Population Institute set up its West Coast office in 1970. That office was under various leaders until Norman Fleishman took over in 1973. Though the organizations lobbying on behalf of population control went through several leadership and structural changes, the issue remained well represented in Hollywood on into the eighties.[3]

From the beginning, population control was a double-edged sword for entertainment television. On the one hand, it lent itself quite easily to dramatization. Topics like teenage pregnancy and abortion were the kinds of "problems" to which social issue drama was naturally drawn. On the other hand, as the experience with Maude's abortion had shown, the incorporation of population control issues into prime time could generate protests from conservative groups. Therefore, the population control groups had to walk a fine line in their dealings with prime-time television.

Throughout the 1970s, the Population Institute maintained a quiet presence in Hollywood. Its primary tactic was to "educate and sensitize." Because of the creative community's resistance to outside pressure, care had to be taken, in Fleishman's words, "to show people I wasn't trying to persuade or propagandize them." Fleishman's trademark became the social gatherings that he held several times a year in private homes. The purpose of these events was to introduce ideas in a subtle way and to stimulate interest. Fleishman referred to this process as "cultivation," tilling the creative soil, planting seeds that may take root and grow.[4]

The Population Institute continued its awards program for several years after the *Maude* controversy, handing out more than $100,000 to writers and producers during the five years it was in operation. The program was discontinued in 1977. According to Fleishman, the Institute stopped the awards because its leaders felt they had established a significant enough presence in Hollywood so that awards were no longer necessary. It was a costly program, and limited funds could be allocated more effectively to other advocacy efforts. The decision could also have had something to do with the rise of conservative groups in the late seventies. Awards increased the Population Institute's visibility, and it was more advantageous to keep a lower profile.[5]

Since it could take a year or two for an idea planted at a meeting to find its way into prime time, it was often difficult to assess the impact of the Population Institute's efforts on programming. But the elusive quality of influence could also work in the organization's favor. Critics of the lobbying group were hard-pressed, for the most part, to identify specific instances of direct involvement with program creation. At the same time, to show his supporters that his work in Hollywood was effective, Fleishman could take credit for certain story lines for which he may or may not have been responsible. "I've seen things happen two years after I met with somebody," Fleishman explained, "and I'll just get a little feeling inside I had something to do with it and I can't prove it."

In his role as a technical consultant, Fleishman's influence was more clearly in evidence. The advocate was frequently called upon to provide statistics, as well as advice, on how an issue could best be translated into drama. Occasionally, this led to rather extensive involvement in the development and writing of scripts. Like the Gay Media Task Force's Newton Deiter, Fleishman offered more than technical assistance. A partnership arrangement developed with a few Hollywood writers and producers. From time to time the lobby group would join forces with producers and writers to do battle with the networks.

Fleishman consulted with Norman Lear on a number of his series. He characterized himself as an important "offstage presence" in an episode of *All in the Family* where Archie's son-in-law Michael had a vasectomy. A few years later, when actress Sally Struthers (who played Michael's wife, Gloria, in the show) became pregnant in real life, CBS wanted to write her pregnancy into the series. The only way that could be done was by making the vasectomy fail. As Fleishman recalled:

> Bob Schiller [the writer] called me and told me that the network wanted this vasectomy to fail. What could we do? I said, "You can't do that. There's no way. They fail so rarely. If you did the show, there'd be fifty million people who saw where a vasectomy failed. You'd send jitters to men all over the country." . . . I thought up all kinds of things they could somehow get by, but for sure you couldn't have the vasectomy fail. Well, we kept talking and kept talking and kept talking. . . . I even got the doctors in the East who were with the Association for Voluntary Sterilization to call Bob and to talk to him, and to write letters to the network.

But before any final decisions were made, Struthers had her baby.

Fleishman also worked with writer Dan Wakefield on ideas for introducing birth control into NBC's *James at 15* series. The show's popularity among teenagers made it an effective vehicle for educational messages. Wakefield and Fleishman brainstormed about subtle ways to incorporate the issue into one of the episodes. One idea was to have a teenage character in the show carrying a wallet with an obvious outline of a condom. Remembered Fleishman: "One of the kids would say, 'I'll give you a dollar,' and another kid says, 'I'll give you two dollars,' and another says, 'I'll give you five dollars.' It was kind of a nice thing to show their respect for contraception." But standards and practices refused to allow the dialogue sequence.[6]

It was an argument with the network over a later episode of *James at 15* that exploded into a public controversy. Fleishman again played a key role. When programming chief Paul Klein suggested that the lead character lose his virginity to a Swedish foreign exchange student on James's sixteenth birthday (which just happened to fall during the important February ratings sweep period), Wakefield agreed to the idea, but insisted that this time a message about birth control be incorporated into the story. The original script included the following exchange of dialogue between the two young people:

James: I love you and I want to protect you. I've heard about teen-age pregnancies and all that and I think people ought to be responsible.
Girl: I am responsible, James.
James: You are? That's great.[7]

Fearing that such a routine reference to contraception might encourage premarital sex among teenage viewers—and worried that conservative groups might protest, network censors insisted on changes in the script. Wakefield refused to comply and resigned in protest. Fleishman helped the writer take his story to the press and held a special honorary event for him to encourage support. As a consequence, the network was subjected to a deluge of angry mail, before and after the controversial broadcast.[8]

In 1980 the Population Institute split into two organizations. The newly created Center for Population Options established its headquarters in Washington, D.C., and also took over the work of lobbying the Hollywood community. Fleishman worked with CPO until 1982, before starting his new organization, Microsecond.

When Marcy Kelly stepped in as the new media director for the Center for Population Options, she too had already established herself within the Hollywood creative community. Her first stint as a Hollywood lobbyist had begun a few years before with the Scott Newman Center, a nonprofit organization set up by actor Paul Newman in the name of his son who had died of a drug overdose in 1978. Like the Population Institute, the Scott Newman Center worked directly with the creative community as well as the networks, to encourage prime-time TV to deal responsibly with the issue of drug abuse.[9]

By the mid-eighties, the Center for Population Options had broadened its scope to encompass not only teenage pregnancy and birth control but also the portrayal of sex roles and sexual behavior in general. Its strategy had also shifted somewhat, from an informal, low-profile involvement to a more institutionalized, visible presence within the Hollywood creative community. CPO formalized its relationships with the television industry by setting up an advisory board comprised of prominent producers and writers as well as network executives. CPO consulted regularly with a number of producers. After-school specials like CBS's *Babies Having Babies* were made with considerable input from CPO, which also helped promote the TV movie. The advocacy group also worked more closely with the network standards and practices departments, enlisting their cooperation along with that of the creative community. With increased funding from several foundations, CPO set up workshops and conferences for creative community members.[10]

CPO revived its awards programs in 1985. This time, instead of offering cash prizes directly to writers and producers, the organization awarded the prizes to universities to encourage research on sexual responsibility and the media.

By that time, there were a number of other awards programs that honored members of the creative community for furthering various causes. Like the Oscars and the Emmys, ceremonies for these awards became yearly gala gatherings, with glamorous Hollywood celebrities as presenters, performers, and recipients. Only a handful of the groups gave out cash prizes as incentives to producers and writers. Most offered plaques and statues and public recognition for contribution to their various causes.

The oldest program was the NAACP's IMAGE awards, which began in 1967. Nosotros patterned its Golden Eagle awards after the IMAGE awards banquet. The Scott Newman Center began awarding cash prizes to the creators of programs dealing with drug and alcohol abuse. The Association of Asian-Pacific American Artists started an awards program to recognize "fully dimensional" portrayal of Asian Americans. The Washington, D.C.-based National Commission on Working Women developed a program to honor outstanding portrayals of working women in prime time.[11]

One of the most influential programs was the Humanitas Award, which was created to "encourage the industry to do audience enrichment in prime time." The criteria for earning the award were purposefully vague, going to writers of scripts that promoted "those values which most fully enrich the human person." By the 1980s substantial awards ranging from $10,000 to $25,000 were being given out yearly to individuals, and the Humanitas had become a kind of quality status symbol in Hollywood. Writers designed certain scripts as Humanitas nominees, consciously directing their efforts toward winning the coveted prize.[12]

There were media awards for so many causes that every possible issue seemed to be covered. Commented one leader of an advocacy group that did not have an awards program: "I wish we could start one, but there's nothing left to give an award for." The awards worked well to give the advocacy groups visibility. The yearly regularity of the awards ceremonies served as reminders that these organizations had established themselves as permanent adjuncts to the entertainment community. A few of the groups succeeded in getting their names listed in the monthly Writers Guild newsletter as part of a regular column of technical consultants on various subjects.

The awards and the consultation services reflected a general trend among a number of advocacy groups toward closer cooperation with the creative community and a more "positive" stance toward the industry in general. This was especially true of the organizations with industry professionals in their membership, who relied on the good will of the television and film industries for their livelihood. Nosotros had been one of the first of these organizations to become established in Hollywood. By the mid-eighties, others had followed the Nosotros model. One of them was the Alliance for Gay and Lesbian Artists (AGLA), which began as a small actors' support group in 1975 and grew to become an active advocacy group with a membership of more than three hundred actors, directors, screenwriters, and other media professionals. AGLA began offering "script consulting services" to screenwriters and producers in 1981. That same year (which was also during the Coalition for Better Television's campaign), AGLA began its own awards program. In 1982, at a well-publicized gala event, AGLA honored actor Tony Randall for his portrayal and creative input in the series *Love, Sidney,* which the Coalition for Better Television had so bitterly attacked.[13]

AGLA provided the same kind of technical consultation service as the Gay Media Task Force, which continued to operate in Hollywood. But the new group had no affiliation with the more militant New York-based National Gay Task Force. AGLA's leaders were careful to characterize their group as one whose interests were mutually compatible with those of other industry professionals. "We at AGLA prefer to be seen as a celebrator, not a watchdog," one of them told the press. "Very often, when gays lobby for change," added another, "we hear negativism and criticism. . . . It is important for us to recognize positive work."[14]

A group with a similar agenda and operating style was the Media Office, which was set up in 1980 to represent the disabled. The Media Office held yearly awards banquets and offered script consulting services to producers and writers. The office also worked directly with standards and practices departments.[15]

Quite often the programs cited for commendation by these advocacy groups had involved extensive input and consultation from the group representatives. Producers and writers were therefore rewarded not only for the way they presented the issues but also for their willingness to cooperate with the organizations promoting those issues in the media. The Media Office's executive director, Tari Susan Hartman, recalled her extensive involvement in the ABC series *The Fall Guy*. In 1984, the producers decided to do an episode about a stunt man who becomes disabled. As Hartman explained: "I met with the writer when he just got assigned the project. . . . He said, 'What should I avoid?' I told him what to avoid. Then everything I warned him about ended up in the script and the script got yanked. We're not sure who yanked it, but we're so well networked, it could have been anyone." Fearing that the script would offend the advocacy group, the producer called Hartman back, to help to "save it." "We met with the producer, the story editor and the director," Hartman recalled. "We reworked almost the entire script. We talked about it line by line, page by page, scene by scene. . . . We even talked about camera angles, ones that were less condescending." Later that year, the Media Office gave the series an award.[16]

Many of the advocacy groups operating in Hollywood found that entertainment television had an uncanny ability to "use up" issues quickly. There were limits to the number of times a particular issue could appear in a series or a season. Programming executives were inclined to say, "We've already done that one," and often the creative community and lobbyists alike were faced with the dilemma of coming up with innovative ways to repackage an issue that television had already treated. One solution to this problem was to incorporate messages into the background of prime time programs. For controversial issues, this strategy could also get around the problem of complaints by opposing groups. There was an added advantage to the strategy. Following the well-known advertising principle of frequent repetition, advocacy groups could reach more people with a long-running show than through a single broadcast. Each group had its own agenda in this area and each experienced varying degrees of success.

AGLA leaders sought increases in the TV characters that "just happened to be gay," rather than those whose homosexuality became the focal point of the drama. This was easier to achieve in the eighties than the seventies, when the whole idea of homosexuality was new to television and to mainstream culture as well. However, the AIDS crisis, which reached epidemic proportions in the gay population by the mid-eighties, made it somewhat difficult to routinely incorporate gay characters into prime time.[17]

The Center for Population Options asked for similar routine references to birth control. As Marcy Kelly explained it: "On any of these shows [in prime time] there will be a moment where two people come together romantically and are going to go to bed together or we see them in bed together. . . . I would like the shows to just take one moment, either a visual moment or a verbal moment, where the woman says 'just a minute, are you protected?' or 'I'm not taking the pill.' . . . It doesn't have to take them 10 seconds to do it. I want to incorporate it as a natural function of daily life. People drink orange juice at breakfast tables. Well, lots of women take a pill every morning." Because the issue was so sensitive, such routine references were not that easy to achieve. However, for CPO, the

AIDs crisis helped draw attention to the need for certain kinds of birth control methods—particularly condoms, which were considered a safeguard against the sexually transmitted disease. As a consequence, Kelly found producers, writers, and networks more willing to incorporate material relating to birth control devices into programs.[18]

Cooperative Consultation—Alcohol Education

The emphasis on the background of prime-time reality was one of the key components of the strategy used by the research team of Warren Breed and James De Foe. With a grant from the U.S. Public Health Service and several foundations, the two "scholar consultants"—as they called themselves—set up a Los Angeles office in 1979 to lobby the production community around the issue of alcohol consumption in entertainment programming. Through a process they referred to as "cooperative consultation" the two engaged in a campaign to change the way drinking was portrayed in prime time. This strategy resembled those employed by other Hollywood lobbyists, but it also had several unique features.[19]

Breed and De Foe first set out to educate and sensitize television industry professionals. Like the other groups, part of the task was to provide statistics and research results that could enlighten creative community members. But, since these efforts were corrective in nature, a more important initial step was to show media professionals what they had been doing wrong. For this purpose, scientific content analyses were conducted to track patterns of drinking in prime time. Many other advocacy groups had used similar studies to point out excesses and inaccuracies in entertainment programming, often to the chagrin of TV industry executives, who did their best to undermine the credibility of the research methods. What Breed and De Foe found, however, was that many members of the creative community were quite interested in their results. One reason was that a number of producers and writers had had personal experiences with alcohol abuse, either themselves or through friends and relatives. Another reason was that the issue was not considered controversial. "While people differ about drinking in general," the two noted, "nobody is in favor of alcoholism or alcohol abuse."[20]

Since James De Foe had worked as a television writer for a number of years, he was able to enlist the help of the Writers Guild as well as the Directors Guild to distribute the results of the content analyses to hundreds of industry people. To raise public awareness of the problem, the study results were also released to the press. The findings were surprising and newsworthy. If viewers were at all influenced by what they saw on television, prime-time programming appeared to be contributing to the problem of alcoholism. With more than eight "drinking acts" per hour in the top sitcoms and dramas (sampled over a three-year period), the report concluded, "alcohol was the most-used beverage on screen." The researchers also found some notable and disturbing patterns, including:

> young people yearning for the time when they too could drink, adults drinking to cope with crisis or stress, alcohol being abused with no visible consequences or reactions by other characters, gratuitous drinking, miraculous recoveries from alcoholism, jokes that excuse immoderate drinking, characters being denied the opportunity to decline a drink, glamorizing drinking, and series regulars receiving less "punishment" than guest characters for alcohol abuse.[21]

Though the studies provided evidence of television's excesses, Breed and De Foe were careful not to lay any blame on the industry for its failure to portray alcohol use responsibly. They assured producers, writers, and directors, for example, that excessive drinking scenes had not been intentional but were merely innocent oversights, the result of unconscious decisions related to the demands of storytelling in a visual medium. "It seemed that drinking scenes were usually not designed as a central part of the story. Rather, they were peripheral, taken for granted, sometimes resulting from the director's imperative to 'move the scene.' Hence, viewers were exposed to symbols not consciously intended by the creative staff." This innocent inclusion of alcohol consumption seemed like a fairly easy problem to remedy. "Gratuitous" drinking could easily be excised without disrupting story lines. Other references could be corrected through a simple change of dialogue or action.[22]

The team held special "sensitivity sessions" with network standards and practices representatives, providing them with guidelines to follow when showing alcohol consumption. "Ten Suggestions for the Portrayal of Alcohol in the Media" proposed the following rules:

1. Do not glamorize the drinking or serving of alcohol as an especially sophisticated or adult pursuit.
2. Do not show the use of alcohol gratuitously—in those cases where another beverage might easily and fittingly be substituted.
3. Do not omit the grim consequences of alcohol misuse or alcoholism.
4. Do not deny characters a chance to refuse an alcoholic drink by statements such as "What will you have?" or "Do you want a drink with the rest of the guys?"
5. Do not show drinking alcohol as an activity which is so "normal" that everyone must indulge; allow for decision making on the part of every character.
6. Do not show excessive drinking without consequences or with only pleasant consequences.
7. Do not show miraculous recoveries from alcoholism; normally, it is a most difficult task.
8. Do not show children "lusting after" alcohol and the time when they are adult enough to drink it.
9. Do not associate drinking alcohol with macho pursuits in such a way that heavy drinking is a requirement for proving oneself as a man.
10. Do not omit the reaction of others to heavy alcoholic drinking especially when it may be a criticism.[23]

The task of approaching individual producers about their own series—a process Breed and De Foe called "specific education"—was not as easy as the "general education" of industry professionals. "Many of the creative personnel never answered our calls," the team recalled. But, through personal contacts a fair number of producers, writers, and directors agreed to cooperate with the researchers. To encourage participation in the project, Breed and De Foe prepared special tailor-made "Episode Reports" of those programs which featured "significant alcohol behavior." These reports provided detailed information on "who drank or talked about drinking on a certain show (over an extended period of time) . . . as well as the context of each incident." "Thus," explained the team, "with Episode Reports dating back some five years on some shows, we actually are more familiar with these shows and their patterns than some writers, directors or producers who may have logged a season or less on that show." Such comprehensive analyses proved to be of interest to program creators. "Most wanted to see how their own work was evaluated by outside observers and what kinds of things they might have done differently. Almost all were surprised by the findings . . . [particularly] that alcohol was the most-used beverage on screen.[24]

By showing producers and writers that they were "doing alcohol education" every day whether they knew it or not, Breed and De Foe found a number of them willing to listen to specific suggestions for changes. Because the two consultants were on hand to help TV creators deal with the alcohol issue in their programs, Breed and De Foe were able to play a more direct role in shaping how that issue was treated. Breed and De Foe did not want to see drinking completely removed from prime time. If television were to play an educational role, it would be more effective to "have drinking placed in an appropriate context so that a realistic picture of alcohol use and abuse could be shown." Taking care not to suggest "overt sermonizing, moralizing, or the injection of ideology," the consultants tried to use their own "knowledge of the program's nuances and/or limitations to offer alternative solutions within the premises of dramatic series." By working with writers and producers, they developed methods for translating their more general goals into specific narrative devices.[25]

Input ranged from the injection of what the consultants called "reality reminders" to full participation in the development of a story line. A "reality reminder" was defined as a "word, phrase, or joke that would point up the reality of drinking without causing anyone to step out of character." A reality reminder was incorporated into one episode of the comedy series *The Jeffersons,* for instance, in the following manner: during a party with lots of drinking, one of the characters, who has gotten himself into a tipsy state, whispers to his wife, 'Let's go upstairs and have some fun." With a knowing look, she answers: "When you're in this condition, the fun is mostly in your imagination."[26]

The consultants were substantially involved in a number of other programs. The writers of *M*A*S*H* asked Breed and De Foe for help during the planning stages of an episode where one of the series' main characters, Hawkeye, decided that he was drinking too much and quit for a week. After some discussion about how best to use the episode to educate viewers, it was decided that another character, Margaret Houlihan, would defend Hawkeye's right to abstain. "Without sounding 'preachy' . . . she could say that 'in this stupid war' people can easily forget they might be drinking too much and too often, and that Hawkeye had a perfect right to stop if he thought he might be losing control over his drinking." A scene was also written at the end of the episode to stress the same point. "After a harrowing incident in the operating room involving a grenade, Hawkeye joins the other surgeons at the bar and a drink is served to him. The other surgeons believe he will drink now, after the crisis. But after deliberating for a few moments, Hawkeye declares, telling his friends, 'I'll have a drink when I want it, not when I need it.' "[27]

One reason why Breed and De Foe were successful at influencing the creative community was that their campaigns occurred during a period when alcohol abuse was becoming a major public issue. Public health and government agencies were devoting energy and resources to the problem, and public interest groups had already mobilized for legislation and greater public awareness.

In 1982, the Caucus of Producers, Writers, and Directors issued a White Paper entitled "We've Done Some Thinking . . ." The document called for formal adoption of the guidelines for the portrayal of alcohol introduced earlier by Breed and De Foe. Caucus leaders explained their decision to draw up the policy paper as a reaction to the recent alcohol-related deaths of entertainment industry celebrities Natalie Wood and William Holden. These events may have provided a dramatic catalyst, but there were other considerations as well. Forces were coalescing to focus further attention on the media.[28]

Public interest groups had already begun pushing the government to ban alcohol advertising from the broadcast airwaves. Tobacco commercials had been banned from radio and TV in 1971, with considerable losses in advertising revenues. In 1983, the Washington-based Center for Science in the Public Interest proposed a similar ban on beer and wine commercials. Within a short time, that group had joined with more than 20 other public interest groups to form a coalition called SMART (Stop Marketing Alcohol on Radio and Television). The broadcasting industry (as well as the liquor industry) responded with the full weight of its political and economic powers. The National Association of Broadcasters set up a special task force to fend off the ban proposal. An Entertainment Industries Council was formed to help the industry "deglamorize drug and alcohol use" and to publicize its efforts in this area. Radio and TV stations began blitzing the airwaves with public service announcements about drunk driving and alcoholism, and ultimately the SMART plan was defeated.[29]

One of the strategies the industry used to ward off the proposed ad restrictions was to stress repeatedly the great strides television had made in correcting the portrayal of alcohol use in prime-time programs. Special videotapes were prepared, with excerpts from various program episodes. These were taken to Washington and entered as testimony during congressional hearings. Industry representatives proudly cited various instances where content had been voluntarily changed. For example:

> A decision was made by the producers of *Matt Houston,* and what had become a "format" drinking scene (with no relationship to the plot) was changed. The "format" scene remains but the beverages are now non-alcoholic. . . . Even though most of the drinking portrayed on *Dallas* (and there was a lot) was in support of the negative images of some of the characters, a recent report from one of the *Dallas* regular directors told us that the use of alcohol in *Dallas* programs has been cut down some seventy percent. "Only when it helps satisfy the text" seems to be the rule now.[30]

The "cooperative consultation" from Breed and De Foe and the industry's subsequent internalized content policies had been successful not only at correcting the portrayal of alcohol but also at protecting the industry from further regulation.

As advocacy groups heard about the successes of Hollywood lobbies, more of them began to follow the lead of these groups. Several East Coast organizations set up Los Angeles offices in the late eighties in order to work more closely with the production community. The National Council for Families and Television had already been holding regular conferences with producers and writers when it decided to establish a West Coast office in 1986. In a monthly "Information Service Bulletin," the group regularly offered TV writers and producers "news and information affecting families and family life." "Our expectation," explained the newsletter, "based on past experience, is that at least some of this material will find expression in television scripts and programs." The AFL-CIO established a West Coast arm of its Labor Institute for Public Affairs in 1984 to encourage better representation of the American worker in prime time.[31]

But not every group that approached Hollywood was entirely successful. If an issue were either too controversial or incompatible with the value system of the creative community, it could be difficult to gain much access and cooperation in Hollywood. Even those organizations with ideas supported by TV producers found that their ability to influence programming was hampered by limitations of the medium. The story of the Solar Lobby's efforts in Hollywood is a case in point.

Solar Lobby

The Solar Lobby, a Washington-based nonprofit group, set up an L.A. office in 1981 and sent twenty-seven-year-old Tyrone Braswell out to California to head it up. The goal was to use entertainment programming to educate the public about the advantages of solar power. With the help of Norman Fleishman, the Solar Lobby modeled its Hollywood operation after that of the Population Institute.

The plan was to get creative community members personally involved with and committed to solar energy. Targeting those producers who already incorporated social themes into their programs, the Solar Lobby planned a series of individual meetings where program ideas tailored to each producer's series would be presented. The plan also called for a "house meeting," where television professionals would hear presentations on solar energy and energy conservation, plus personal consultation at the homes of interested producers about how their own houses might be converted to solar energy.[32]

But Braswell found that, though a number of producers were open to hearing about solar energy, only a very few of them agreed to incorporate the issue into their programs. There were a few successes: a solar house on the popular daytime soap opera, *All My Children;* a news story about solar energy on *Lou Grant;* and a solar-powered still on *The Dukes of Hazzard*. However, for the most part, the Solar Lobby was able to get only brief plugs and passing references to solar energy into entertainment TV. In some cases, the lobby agreed not to take credit for the script changes. This put Braswell in a difficult position, because members of the Solar Lobby who were not familiar with the sensitivities of dealing with entertainment television wanted more visible proof that their efforts were paying off, and they wanted to publicize those successes for fundraising purposes. When Braswell reported his accomplishments to the general membership, he often had to explain the significance of the minor pieces of dialogue that he had negotiated with producers to include. The advocate managed to get a line in *Dallas* where Bobby Ewing asked his father to invest in solar energy, though the father never went along with the idea. Asked by a reporter whether this put solar in a bad light, Braswell enthusiastically responded: "Oh no, it was terrific. Because it was Bobby Ewing saying that solar is the wave of the future, that they've got to get into solar. This was great—the young telling the old that solar is the wave of the future. It was right-on."[33]

But these small success stories were not enough to keep the Solar Lobby alive in L.A., and after eighteen months of operation, the project was discontinued. Though a number of factors—including internal organizational problems—contributed to the office's demise, there were inherent problems with the issue. For one thing, it was difficult to make solar energy a compelling dramatic element in entertainment programming. Topics such as unwanted pregnancy, alcoholism, drug abuse, and homosexuality presented many possibilities for narrative treatment, but, as Braswell himself admitted, "the problem with solar energy is that it's basically boring; it's like talking about your refrigerator."[34]

The other problem was that the solar power issue didn't have much staying power. For a short time during the late seventies and early eighties when much public attention focused on the energy crisis, solar energy was appealing to Hollywood as a trendy and popular issue that could, even in small doses, lend credibility and timeliness to TV programs. But, as the popularity of the issue faded, so did its usefulness to network television. Braswell used a Southern California metaphor to describe his short-lived experience in Hollywood: "I was," he said, "like a surfer, who is only as good as the wave coming in."[35]

CHAPTER TEN

Packaging Controversy

The real world and the TV world were mirroring each other. While half a dozen picketers marched along the sidewalk in front of a Santa Monica movie house, the audience inside watched a similar scene on the screen. In an upcoming episode of the popular CBS cop show, *Cagney & Lacey,* right-to-life activists were picketing an abortion clinic.

This special pre-screening took place one week before the episode's scheduled broadcast on November 11, 1985. Many of those in the attendance were members of the National Organization for Women, who had been invited by the show's producer. The protesters outside the theater were from local right-to-life organizations. They had not been invited to the screening, but had heard about it through the press, where leaders of the National Right to Life Committee were publicly urging CBS to cancel the controversial episode.[1]

At first glance, it looked like the *Maude* controversy all over again. But this case was markedly different. The protest over *Cagney & Lacey* did not spring up spontaneously after the broadcast. It was part of a publicity ploy created by *Cagney & Lacey*'s executive producer. The network's handling of the episode, as well as the way in which the controversy was generated and resolved, revealed a number of changes that had taken place in network television in the thirteen years since *Maude*'s stormy season debut.

By 1985, controversial issues were part of the landscape of entertainment television. The trend begun in the early seventies had steadily continued. Though the campaign by the religious right had put a temporary damper on provocative content a few years earlier, as soon as the pressure had subsided, issue-oriented programs had begun to find their way back into the prime-time schedule. The religious right had not disappeared; conservative groups still occasionally attacked programs, using their standard technique of blitzing all possible advertisers with protest letters. In a few instances, such a blanket approach successfully scared off some sponsors. After such a campaign in 1983, a dozen companies pulled out of NBC's made-for-TV movie, *Sessions,* which starred Veronica Hamel as a prostitute. But, for the most part, the network television industry had proven remarkably adaptive and resilient in deflecting such pressures.[2]

The watershed case had been ABC's 1983 nuclear holocaust movie, *The Day After.* Letter-writing campaigns by conservatives had prompted most of the major advertisers to withdraw their ads from the movie, forcing ABC to sell its thirty-second spots at distress rates. This made the movie more affordable for smaller advertisers who would generally never be able to participate in blockbuster programming. The advertisers that had dropped out of the show feared that viewers would make a negative association between their products and the controversial program, with it downbeat theme. Given the earlier experiences with the anti-violence and Moral Majority campaigns, these fears seemed justified. But the J. Walter Thompson Company challenged this assumption with a special survey of viewers. (This was the same ad agency whose mid-seventies studies of consumer attitudes toward the sponsors of violent content had so alarmed the industry.) The results of the new study were both surprising and encouraging. Advertisers were not only unharmed by their participation in the controversial broadcast, but their association with *The Day After* actually enhanced their image. Most viewers, the study showed, "came away from *The Day After* with attitudes that were unchanged toward the sponsors. In cases where there were attitude shifts, the positive shifts outweighed the negative by a margin of about 5 to 1."[3]

This new research reflected continuing efforts by the advertising industry to define more clearly the rather murky area of viewer association between products and program content. Of course there were some flaws in this study. *The Day After* was not exactly a typical case. Though it had been hotly debated by political groups on the right and the left, the promotional effort behind the film had turned it into a national event of major proportions, hardly comparable to a TV movie like *Sessions*. But the important point was that the television industry was calling the bluff on boycott threats. ABC's social research department later followed up on the J. Walter Thompson survey with its own study of controversial programming. Confirming the earlier findings, the new report concluded that "viewers do not hold advertisers responsible for a program's content." These studies were designed to persuade advertisers that buying time in controversial programs was a safe enterprise. The added benefit of such programs was that they generated higher ratings than the more conventional shows. *The Day After* had garnered 62 percent of the viewing audience, making it the twelfth-highest rated program ever aired on TV.[4]

ABC's success with *The Day After* helped pave the way for more and more daring material in prime time. This renewed interest in controversy was also part of the network strategy to win back the audiences that were being lured away by the more provocative programming on cable television and in video stores. Just as controversial social issues successfully attracted the desirably demographic groups in the seventies, it was now viewed as the ingredient to draw audiences back to network television in the face of competition. Explained Perry Lafferty, a senior vice president at NBC: "How do the networks fight back against cable? We can't do it by putting on more violence and sex, but we can probe social issues that haven's been explored." Network television's "most effective weapon"—its "big gun," according to the executive—was the social issue movie.[5]

The controversial social issue movie, already a staple in prime-time television, now emerged as a highly developed formula, which successfully packaged several key ingredients to ensure high ratings. The most important element was a high-profile issue or story, one that viewers were already familiar with through newspapers, magazines, TV news and talk shows. Because they were already so visible in the mass media, these "headline stories" had a built-in draw. Some were taken directly from the real event. "Based-ons"—as industry insiders labeled them—promised to bring real characters to life in American living rooms, where they could tell their own dramatic stories.

The movies were well researched, involving extensive consultation with special interest advocacy groups and professionals from the medical, social welfare, and legal communities. Indeed, they appeared to be almost documentary in nature. The new brand of TV movie also carried a certain amount of prestige with it, which created a positive environment for advertisers. As *The Day After* research had shown, even downbeat content could be counterbalanced by the prestige factor.

Above all, these movies were "promotable." Promotability was an essential ingredient, because television movies came and went so quickly. Unlike theatrical films, which had time to build audiences through advertising, critical reviews, and word of mouth, TV movies lived a short life of two or three hours on one evening. Many of the movies were scheduled during the ratings sweep periods, and the networks went to great lengths to publicize them. In addition to the standard publicity efforts—ads in *TV Guide* and newspapers, on-air promotional spots, press screenings, and so on—the networks had developed elaborate techniques for stimulating viewer interest before a broadcast. Stars of upcoming TV movies, sometimes accompanied by experts, appeared as guests on morning talk shows discussing the importance of the movie's issue. Special screenings were sometimes arranged for advocacy groups. And the networks frequently used the New York-based Cultural Information Service (CIS) to promote TV movies to the schools. Run by a husband-and-wife team out of their Manhattan loft, this nonprofit organization prepared special study guides, discussion questions, and posters for use with upcoming movies. Weeks before the broadcast, these materials would be mailed to schools around the country to encourage teachers to assign the programs to their students and then talk about them in class. Materials were also mailed out to the network affiliates, whose news departments would sometimes coordinate their local newscasts with the TV movie.[6]

During the 1984–85 season, social issue movies were particularly visible in prime time. ABC's *Something About Amelia* dramatized the poignant and painful experience of incest between a father and his daughter. On the same network, Laura Z. Hobson's well-known novel *Consenting Adults,* about a young gay man's decision to come out of the closet, was adapted for television. The producer and network worked with the Alliance for Gay and Lesbian Artists during the writing of the movie and pre-screened it in advance to gay groups. Representatives of the National Organization for Women worked with the producers of *The Burning Bed* "every step of the way" during its writing and production. This NBC movie was based on the true story of an abused wife who murdered her husband by setting his bed on fire. Also broadcast that year was *The Atlanta Child Murders,* a CBS film based on the mutilation-killings of twenty-eight young people in Atlanta a few years earlier.[7]

Controversial social issues were also a boon for series television. When tied to promotional efforts, a special episode of a situation comedy could boost ratings considerably. When the producers of *Webster* did a special episode on child molestation in late January, the show quickly climbed into the top ten for the first time all season. Referring to the issue-oriented sitcom as a "major prime-time phenomenon," *USA Today* cited a number of other examples that year:

> Arnold hits the bottle this Saturday on NBC's *Diff'rent Strokes.* They'll play nuclear war games on ABC's *Benson.* And upcoming episodes of *Silver Spoons* and *Family Ties,* both on NBC, will address dyslexia and parental death.

If a series had been on the air for some time, these special episodes could restimulate viewer interest in the weekly shows, especially if an article highlighting the program appeared in the paper the day before the broadcast. "Most newspapers won't pay attention to a show after its second year," explained one studio executive, "unless you do something different."[8]

The packaging and promotion of controversy in prime time

was never done without the full participation of the network standards and practices departments. Their role in the process was essential. Responsible for managing political advocacy groups as well as overseeing program content, standards and practices executives had become the network specialists in political controversy. They had developed special antennae for sensing trouble. They were uniquely skilled at spotting a word, a phrase, or a plot sequence that could evoke a negative reaction from any one of the two hundred or so groups that had made themselves known to the networks. They also knew which topics required the most sensitive treatment.

Sensitive issues were not confined to those subjects which involved social taboos or sexually explicit references. An issue was also considered "sensitive" if there was an advocacy group, known to the networks, with an active interest in how that issue was presented on television. Not only were abortion, homosexuality, and gun control "sensitive," but so were seemingly innocuous issues which, because of advocacy groups, had become "charged." Even the potato could be "sensitive," according to one network executive who told the story of how a trade group for the potato industry complained when a TV character turned down a helping of the vegetable because she was on a diet. Of course, some issues were more sensitive than others, depending on the power and size of the group, the nature of its relationship with the network, and the potential for trouble.

The job of the standards and practices departments was not to eliminate sensitive issues but to find ways to structure them into entertainment programming so they would stimulate audience interest without evoking protests. This was done through a set of content policies which, like the strategies for managing advocacy groups, served to defuse pressure. These policies were finely tuned mechanisms which functioned to keep various forces—both outside and within the television industry—in balance. Although they produced fairly consistent patterns in the treatment of sensitive issues, they were dynamic and adaptive, capable of changing in response to shifts in the behavior of certain groups.[9]

Established routines were employed for controlling the content of entertainment programming on a regular basis. An "editor" was assigned to each series or TV movie. At every step throughout the development, writing, and production of the program, the editor operated as a liaison with the producer. The more controversial the material, the more closely involved the standards and practices department would be.[10]

Though the networks had written content policies, these were purposely vague and general. Like the NAB Code, they served primarily a public relations function. (The NAB Code itself had been voluntarily eliminated in 1982, after a Justice Department anti-trust decree.) The functional policies for handling sensitive issues were not really codified. They evolved over time, worked out on a case-by-case basis through a combination of negotiation and conflict. Sometimes they resulted from the direct input of an advocacy group acting as a consultant during the production of a television show. More often they came out of a negative reaction from one of the groups. Each of these experiences became an important "lesson" which the standards and practices department regularly passed along to its own staff and periodically communicated to the creative community. Through trial and error, certain ways of treating issues were found that maintained the delicate balance between outside and inside pressures. Once a specific approach was used successfully, it would be employed again, this time with much less discussion. In time, the aggregation of these specific, individual decisions became standard operating procedure.[11]

Because the standards and practices editors performed such a critical and delicate role, the creative community learned to accept the network rules for handling sensitive material. A few producers might complain occasionally about the intrusiveness of the censors, but most would agree with the comments of this producer:

> I look at it this way: there are certain rules in writing for television, certain preconditions. It's not any good to complain about them. This is reality. . . . They [the standards and practices departments] are often right from their point of view. They might not be right objectively in what would be better for the show, but they're right from the amount of trouble—"use this line and this is what is going to happen."

Ultimately, the policies for dealing with sensitive issues became internalized by all the key participants in the program decision-making process.[12]

Once a series had been on the air for a while, relations between the producers and the standards and practices department became more routine. Editors who had worked for a long time with specific series generally knew what to expect from producers, and vice versa. It was often easier to inject controversial elements into an episodic program than into a one-time movie, since series generally had loyal audiences, who were familiar with the characters and had a set of expectations about what would appear in the weekly shows. Nevertheless there were still some issues that, no matter where they appeared, required especially careful treatment. Abortion was one of them. The experience with Maude's abortion had been a sobering one for network television. Consequently, that issue was hardly touched in prime-time programming for years afterward. Especially after the rise of the Right to Life movement, the abortion issue was considered one of the most controversial and explosive issues entertainment television could deal with. The few times when it had been broached, standards and practices departments were heavily involved.

By the time *Cagney & Lacey*'s producers decided to do an episode about abortion, the show had been running on CBS for three years. The series featured two female police detectives working together in New York. Its first season had been rough. Only because of the extraordinary efforts of executive producer Barney Rosenzweig had the show survived. After the network had cancelled the series twice, Rosenzweig—who started in the business as a publicist—made the unprecedented move of going directly to the National Organization for Women and orchestrating his own massive letter-writing campaign to convince CBS to reinstate the show. Since then, the series had won several Emmys and had gradually built a strong audience, a large proportion of which was female, urban, upwardly mobile, and college-educated.[13]

Cagney & Lacey had already dealt with some pretty strong issues, including breast cancer and alcoholism, when the producers came up with the abortion idea. The plot featured the two policewomen tracking down the person responsible for bombing an abortion clinic. Bomb threats and actual bombings of abortion clinics around the country had been in the news recently, so the idea was good topical material. The plan for the episode was also to reveal that one of the show's main characters, Mary Beth Lacey—who at the time was expecting a baby—had once had an illegal abortion.[14]

Given the feminist orientation of the series and the show's following, abortion was an appropriate topic. But both the series creators and the network already knew that the issue would require extremely careful handling. Christopher Davidson, the standards and practices editor assigned to the *Cagney & Lacey* series, remembers getting a phone call from the show's co-producer, Steve Brown, a year before the episode aired. "We're thinking of doing an episode about abortion. What do you think about that?" the network executive was asked. Davidson was not particularly pleased with this news, because he knew it would mean a lot of extra work for the standards and practices department, and could cause trouble. His first impulse was to say, "Please don't do that." But he knew better; like so many other issues, abortion was no longer taboo. Davidson did warn the producer that he was "heading into some very troubled territory." There were rules that would have to be followed. "The first words out my mouth," the executive recalls, were that "you will have to do it from a balanced point of view. . . . Look for as much balance as possible."[15]

By the mid-eighties, "balance" had become the most pervasive of the network policies for dealing with controversial material. Since its half-hearted application in the *Maude* abortion episode, the policy had evolved into a well-established convention in entertainment television.

Balance required that controversial material be carefully structured so that network entertainment programs would not appear to be advocating a particular point of view. The policy was tied to the FCC's Fairness Doctrine, which various advocacy groups had tried to use as a weapon against the networks. As a general strategy, it was much easier for the networks to avoid a Fairness complaint than to spend time and trouble

Scene from CBS TV series *Cagney & Lacey*. *(Courtesy of Orion Television)*

fighting one. As a consequence, balance had become standard operating procedure in prime time. The FCC had never interpreted the Fairness Doctrine to mean that each program must be balanced. The Commission and the courts had consistently ruled that balance applied only to overall programming. As the *Maude* case had shown, this meant that one point of view could be expressed in one program as long as other points of view were represented elsewhere in the broadcast schedule. As a pre-emptive policy, however, it was much less risky to have balance within each program. In this way viewers were left with no doubt that the network had fulfilled any possible Fairness Doctrine obligations.[16]

Balance also served a marketing function. Since controversial issues could be divisive, a program that strongly asserted one point of view might alienate audience members who disagreed with that position. Offering various points of view could broaden a program's appeal. And by not taking a position, it could protect advertisers from pressure as well.

As a content policy, balance could take a number of forms. If the sensitive issue was racial stereotyping, a negative character could be balanced by the introduction of a positive character. The standards and practices department had advised producers of ABC's off-beat comedy series *Soap* that "in order to be able to treat the Mafia storyline here and throughout, it will be necessary to introduce a principle continuing character of Italian descent who is very positive and who will, through the dialogue and action, balance and counter any negative stereotypes." Balance also meant that if a show had a black criminal, it would also have a black cop. In programs where the controversial issue formed the core of the story line, balance might be achieved by incorporating a subplot to counterpoint the perspective of the main story.[17]

Even a brief reference to a controversial issue required special dialogue or characters to counterbalance it. One editor reported flagging a sequence about gun control in a script. The characters in a weekly cop show seemed to be taking the position that teflon-coated bullets should be outlawed. Because the powerful National Rifle Association had opposed legislation banning the controversial bullet, and, more importantly, since the gun lobby had attacked network documentaries as well as entertainment programs in the past, the editor insisted that a line of dialogue be inserted which said that "there are people who believe that legislation against weapons is an infringement of their rights." The producer complied with the directive, but, as the editor remembers, when the film was shot, the character read the line in such a way that it looked like he was ridiculing people who opposed anti-gun legislation. The standards and practices department was sufficiently alarmed about the way the line sounded that network executives considered making the producer reshoot the scene. Since this would have been costly and time-consuming, standards and practices checked with the network's legal department to determine "if the news division had done any stories lately that might have allowed the NRA point of view to be heard." When standards and practices executives were assured that this had been done, the department allowed the line to be broadcast without any further changes.[18]

This kind of deliberate, detailed treatment had become commonplace in prime-time television. Some people complained about it, but most members of the creative community accepted the fact that, when dealing with a controversial issue, balance went with the territory. How balance was executed in a particular program, however, was a matter of negotiation.

In the *Cagney & Lacey* episode, balance was worked out through a series of discussions (over several months) among the producers, the writers, and network executives. The various participants in this process each had their own goals, objectives, and interests. Compromises and adjustments had to be made. All of these helped shape the way the issue of abortion was translated into narrative elements.

The standards and practices department preferred the purest form of balance. This meant having the two female cops taking opposing points of view on the issue. "The initial dream from a policy standpoint," remembers Davidson, "was that Christine Cagney [who had been raised Catholic] would at least lean and have a lot of sympathy to the pro-life, and that [the character played by] Tyne Daly [Mary Beth Lacey] would be the pro-choice person . . . that would be terrific." Barney Rosenzweig recalled the first discussion differently. The network executives did not just *suggest* this form of balance, they *demanded* it. "They said to me," he recalls, " 'one of your leads *must* be pro-choice, and one must be pro-life'." The producers were not willing to accept this plan. Their primary argument was that it would be inconsistent with the characters in the series and the audience's expectations about those characters. Rosenzweig finally convinced standards and practices that such a neat form of balance, though desirable for legal reasons, would not work dramatically.[19]

If they could not use the two female leads for balance, then the next best thing, according to standards and practices, would be to have Mary Beth's husband, Harvey, take an anti-abortion stance. But again, this idea did not make much sense dramatically. "We all decided," remembers Davidson, "that it would be out of character for them [Harvey and Mary Beth] to fight about this issue. They had fights, but on philosophical issues he is always supportive and that is part of their charm as a couple."[20]

These problems reflected some of the constraints encountered when working with a series as opposed to a made-for-TV movie. While the movies that had dealt with abortion had balance structured into the basic narrative of the films, an episodic series could not be as easily molded to fit the balance requirements. Davidson understood the limitations of this particular series. Having worked with the show since it began, he knew who the characters were and what you could and could not do with them. But, there were other devices. One of them was to use "peripherals." This included minor regular characters, like Christine Cagney's father, an Irish Catholic blue-collar worker, a perfect spokesman for anti-abortion sentiments. In addition, a special character was written into this particular episode. Added to the story was a female right-to-life activist, whose principal function was to articulate, with conviction, the arguments against abortion.[21]

The standards and practices department did not consult with, nor insist that the producers consult with representatives from the right-to-life movement, even though its political forces were strong and visible. Executives at CBS believed they had enough experience with this issue and enough familiarity with the arguments surrounding it that such consultation was unnecessary. Because there was a "diversity of people working within the standards and practices department at CBS," Davidson explained, the network had sufficient contact with people supporting a right-to-life point of view. The network may also have preferred not to deal directly with anti-abortion activists because it would just be too much trouble. It was more effective, they had learned over the years, to internalize the views of some groups and incorporate them into the programming as an anticipatory strategy to prevent criticism.[22]

Though the standards and practices executives were fairly confident that this *Cagney & Lacey* episode was being handled safely, they did not want to take any chances with it. Their job was to protect the network from external criticism, and they wanted to be sure that this show was not going to get them into trouble, either with outsiders or with their own corporate management. When the script came in from the production company, it was sent to the top brass at CBS for review. This was done far enough in advance of the broadcast, so that, as Davidson put it, "there would be lots of time to argue." The answer came back from management that the show was "not balanced enough." As Davidson remembers: "The structure of balance was there. But it wasn't fleshed out." So, the standards and practices executives continued working with the show's producers to strengthen and "fine tune" the balance. In addition to amplifying the speech by Cagney's father, the dialogue from the pro-life leader was expanded and made more forceful. The network also made suggestions about casting for this role. Remembers Rosenzweig, "We were asked to be sure and cast an attractive woman, not an old grey-haired lady, a blue-haired lady with tennis shoes." What the network really wanted, in order to make the character as strong as possible, was to cast a star in the role. Recalls Davidson: "We were sitting here, saying, 'Boy, wouldn't it be great if they could get a superstar to do this as a cameo, and really do it up.' " Though the producers didn't choose a "superstar," they did cast a well-known actress, Fionnula Flanagan, in the role of the pro-life leader. As Rosenzweig described her, this actress was an "attractive, interesting, mature, grown-up, very sexy lady. . . . We had her say her position clearly, intelligently, with a minimum of emotion, [indicating] that this was an intellectual, political, and emotional decision on her part. . . . She was not frazzled, she was not a crazed fanatic."[23]

While it had been agreed that Cagney and Lacey would not take opposing views on the abortion issue, the network did urge the producers to "drive as much of a wedge as possible between the two characters." Davidson's justification was partly based on the argument that such a change would add more "dramatic conflict" to the episode:

> The earlier script was less interesting because there was no conflict, really. Cagney and Lacey were more or less in agreement that this was a woman's right, it was their job as policewomen to break up this demonstration because the victim lady couldn't get through and she was in an unfortunate situation. You looked at it and said, "Well, this is boring."[24]

But what he really meant was that the show still was not balanced enough. One way to correct the problem, remembered the network executive, was "to put as much of Christine on the fence as we could." Adjustments were made, and with the next drafts of the script, the standards and practices executives felt they had "lurched another step closer." But, recalls Davidson, "it needed a few more lines here, a few more lines there. Maybe tone this speech down . . . a little bit more on the pro-life lady, 'Christine, would you please disagree a little bit more with Tyne here,' and boy were we glad for that scene at the end with Charlie in which he says, 'No way, a Catholic, you just don't get involved with this abortion stuff. It's a sin.' "[25]

By the time the script had gone through this whole process, it was, in Davidson's words, "surprisingly balanced. . . . We got the word back from corporate that they were pleased with the way we had handled the show, and they felt they could stand behind it." But though they had participated heavily throughout the writing of this episode, standards and practices executives still had to wait until the episode had been filmed and edited before determining whether or not it was sufficiently balanced. They knew from experience that what was agreed upon in a script could look quite different on the screen. Final approval would wait until standards and practices had screened the "rough cut" (the show in its finally assembled but not yet polished form). Rosenzweig assured the network executives he would present them with an acceptable episode. "Thirty percent of what you guys want," he promised, "will be in the performances."[26]

Though Chris Davidson felt quite confident about the episode, he approached the rough cut screening with some apprehension. "My boss was on vacation," Davidson recalled, "so I was kind of on the hot seat because I had been in all these meetings and it was my problem." The screening was well attended. The executive remembered the scene in some detail:

> It was a particularly full room . . . a plush room down in the basement. . . . You typically have a whole seating arrangement based on the pecking order—the gofers and the little assistants to the producer sit in front, then the cast, then standards and practices, then programming executives. Then in the very, very back row would be the president of the entertainment division or the vice-president of programs and the executive producer. The writers would be somewhere in the middle. Of course, the main chair is the one that controls the volume, that's where the vice president for programs sits.

Davidson nervously sat through the one-hour program, not knowing exactly what to expect when it ended. He was surprised and pleased at the reaction. As he later described it:

> Usually at those screenings, when the lights come up, everybody looks at the vice president in charge of programming [at the time Harvey Shepard]. Then he'll either say, "It's a charming show" or "I didn't understand why this" or "fix that," etc. But, for the first time in my career, the lights came up and everybody looked at me—including Harvey. I mean everybody. And it was like, "Well?" I just looked at Barney and I said, "You gave us everything we asked for and more. . . . You've done it and you've done it splendidly and it's beautiful and we can salute it."[27]

Even though this *Cagney & Lacey* episode was scheduled during the November ratings sweep period, and the network planned to do some special promotional spots to draw attention to it, it is doubtful whether any controversy would have been generated. That same season, the abortion issue was also treated in two other network crime dramas, *Spenser for Hire* and *Helltown,* and there were no major outcries from right-to-life groups. The on-air promotional spots for *Cagney & Lacey* that CBS was planning to run would be shown so close to the broadcast that opposing groups would have little time to organize a campaign around the show. The episode would be on and off before anything had time to happen. More importantly, those who watched the show would be disarmed by the way the issue had been treated. In contrast to the controversial *Maude* episode, this program was unlikely to elicit a major protest.[28]

It is true that *Cagney & Lacey* had encountered some difficulty with right-wing groups in the past, but the circumstances had been highly unusual. In 1982, CBS had pulled an episode off the air within hours of its scheduled broadcast because of pressures on affiliates. The show featured the two feminist cops protecting an anti-Equal Rights Amendment activist—with remarkable similarity to Phyllis Schlafly—from a psychopathic killer. The network postponed the episode because it was scheduled to air during the final week of debate and voting in Congress on the ERA. CBS decided it would be wiser to postpone the program until after Congress had reached its decision than to risk protest from political groups by airing it during the crucial voting period. Of course, the program had undergone the same kind of careful treatment that the later abortion clinic episode had received. As one reporter noted, "the show was so balanced and apolitical that it was unclear whether the request for equal time would come from pro- or anti-[ERA] forces."[29]

Though Rosenzweig told the press he feared another, similar incident, he did not really believe the right-to-life movement was strong enough, large enough, or organized enough to pose a threat at this point in time. "I was operating with the belief that this was old news. . . . Standards and practices is nervous—of course they're nervous. There's nine little old biddies who are going to write them letters. Well, they don't like to get nine letters. . . . But, there's no issue here. That was settled in 1973." He did believe, however, that he might be able to "create a controversy where none exists" by launching his own publicity campaign far enough in advance of the broadcast so that right-to-life activists would get wind of it. "I tried to create opposition," he recalls, "because I wanted the publicity. I wanted the promotion." Any efforts by right-wing groups to pressure the network could then work to the show's advantage. "I thought I could make a *cause celebre* out of it," he adds, "so that anybody who tries to censor me now, does so at their own peril."[30]

The *Cagney & Lacey* show already had a special relationship with women's groups. There were a number of women working as writers, producers, and story editors on the show, and some were active in political groups such as the National Organization for Women (NOW) and the National Abortion Rights Action League (NARAL). These groups had rallied around the show a few years earlier, when Rosenzweig had orchestrated his "save the show" letter-writing campaign. At that time, the *Cagney & Lacey* producers had shown their support and appreciation by holding fundraisers for organizations. This alliance between the Hollywood producer and the advocacy group was advantageous for both sides. The leaders of NOW—a group which had earlier taken a militant stance against network television, filing petitions to deny station licenses and publishing scathing reports of prime time's misrepresentation of women—had decided they could accomplish more by working with selected producers like Rosenzweig on the shows the group believed portrayed women and women's issues favorably. In addition, the National Commission on Working Women had given the show one of its "Alice" awards.[31]

To launch his campaign, Rosenzweig flew to Washington, D.C., and set up press conferences with several national women's groups, warning them about the expected backlash from right-to-life organizations. Though he had complied with most of the network requirements for balance, the producer still believed the episode supported the pro-choice position, and he publicly took credit for getting that point of view across in the show.[32]

In late summer of 1985, NOW issued an "Action Alert" to its

members, bracing them for trouble. "On November 11, 1985," the bulletin read, "CBS will broadcast a *Cagney & Lacey* episode entitled 'The Clinic.' The content of the show addresses the issue of abortion and the anti-abortion terrorist bombing of clinics and has Cagney and Lacey coming out pro-choice. We anticipate problems from the anti-abortion forces, as this show airs during sweeps week. . . . and will be highly publicized. It is a possibility the anti-abortion forces could try to suppress the network or their affiliates from airing the show." In an action plan reminiscent of the *Maude* campaign, NOW members were urged to write letters to the production company, the network, and the local affiliates. Further strategies were also being considered, the action alert explained, including "the possibility of doing an offensive action by writing the affiliates prior to the November 11th air date and telling them how much we are looking forward to watching the show." Individual chapters were instructed to send for a videotape so that members could "show a brief preview of the episode at a general membership or council meeting prior to November."[33]

Pre-screening to interest groups was a fairly common practice by now. In fact, at this time, NBC was holding a series of screenings for *An Early Frost*, the made-for-TV movie about AIDS that it had scheduled in the time slot opposite the *Cagney & Lacey* abortion clinic program. Special events were being held in New York, Washington, D.C., San Francisco, and Los Angeles, where gay groups and other community organizations were invited to watch the movie and hear public health experts discuss the issue.

One of the unique characteristics of the *Cagney & Lacey* campaign, however, was that, in addition to formal screenings to large audiences, the *Cagney & Lacey* episode was being sent around the country on videocassettes for use in locally arranged grassroots screenings. Though not nearly as large in scale, this pre-broadcast campaign used similar techniques to those employed by nuclear-freeze activists in *The Day After* promotional effort. Because of the advent of low-cost, accessible videotape technology, the program itself could be duplicated and distributed almost as easily as a written memo.

As the press began to publicize the *Cagney & Lacey* controversy, CBS management became more and more alarmed. Rosenzweig remembers getting several calls from very agitated high-level network executives. "Barney, what are you doing?" they asked him. Though they had no real control over what the producer told the press, they tried to get him to characterize the episode in less inflammatory terms. They were particularly upset by Rosenzweig's public statements that the show was pro-choice. "Look, Barney," they told him, "we've taken a look at this show and we believe it's balanced. Will you please say that in your interviews?" The producer agreed to start using the word "balance" when describing the episode, even though he still maintained it was not entirely balanced.[34]

But right-to-life activists were already well aware of the publicity campaign and had taken special note of the statements Rosenzweig and others had made to the press about the show. Daniel Donehey, public relations director of the Washington-based National Right to Life Committee (NRL), was particularly concerned with Barney Rosenzweig's apparent waffling on the show's point of view. Donehey had also read an earlier article in the *New York Times Magazine* discussing how the producers planned to deal with actress Tyne Daly's real-life pregnancy. The piece included a discussion among the writers and producers about the pro-choice orientation of the show and their commitment not to allow it to advocate an anti-abortion point of view. Though right-to-life activists had not been invited to any of the producer-sponsored screenings, a few of their members had managed to attend, and had heard actress Fionnula Flanagan—who played the part of the pro-life activist in the show—explaining that she herself was strongly pro-choice.[35]

The National Right to Life Committee had not established routine contact with the standards and practices departments at the three networks, though the group's leaders had complained periodically about shows and had been invited to a few screenings. ABC had included NRL leaders along with other right-to-life advocates and pro-choice leaders in its pre-broadcast screening of the TV movie *Choices* a few months earlier. NBC executives had met with NRL leaders in 1983 after complaints about two episodes of *Buffalo Bill* which had dealt with abortion. At that time, Donehey recalls, NBC executives suggested that the National Right to Life Committee set up a Hollywood office to lobby producers. No such action had been taken. It is doubtful whether such a direct approach to Hollywood on this issue would have worked anyway.[36]

Unable to get a tape of the show from the producers, Donehey phoned CBS headquarters in New York and demanded to be shown the episode in advance. CBS executives agreed to let him see it. When they finally viewed the program, right-to-life leaders could see that care had been taken to balance the treatment of the issue. But in their opinion, the balance was inadequate. Their interpretation of the program was strongly influenced by what they had been hearing in the press. Though the Fionnula Flanagan character played a substantial role in the story, Donehey believed the actress's personal beliefs showed through her performance, making the character a "strident, unsympathetic stereotype," someone "very hard to identify with and relate to."[37]

With only a few days remaining before the scheduled air date, the hastily launched protest of the right to life groups played directly into the hands of the show's publicity campaign. "The National Right to Life Committee," read the story on the Associated Press wire, "has asked CBS not to air Monday's episode of *Cagney & Lacey* because, it says, the episode about the bombing of an abortion clinic is 'a piece of pure political propaganda' promoting abortion."[38]

The protest not only made newspapers but also was covered as a news story on CBS television. When right-to-life members picketed the private screening of the episode in Los Angeles a few days before it aired, local KCBS news crews were there to catch it on tape and run it on the 11 o'clock news later that night. (Curiously, none of the other local stations thought the incident newsworthy enough to cover.) On the day the episode was scheduled for broadcast, Daniel Donehey and Barney Rosenzweig debated with each other on the *CBS Morning News*.[39]

While CBS was covering the *Cagney & Lacey* controversy, NBC was actively engaged in its own promotional campaign for *An Early Frost*. AIDS experts were appearing on NBC network news and talk shows, and local stations were planning features on their regular news broadcasts about the epidemic. News breaks were being written to interrupt the movie and urge viewers to stay tuned to local news for a follow-up on the issue. NBC also scheduled a special panel discussion to follow its highly promoted broadcast.

Observed TV critic Howard Rosenberg: "Well, AIDS is sizzling right now. With good reason, then, Rosenzweig is worried that 'The Clinic' will get lost in NBC's promotion blitz for *An Early Frost*. Apparently so worried, in fact, that he appears to have launched a promotion campaign of his own, spinning controversy from straw. After all, a little tiff couldn't hurt the box office. You can't blame Rosenzweig for trying to start a drum roll, even if it is self serving. . . . But what criticism? What pressure? Before R. spoke out there was none, unless it was expressed in whispers. There certainly was no public threat by a pro-life group to cut down 'The Clinic' before it could air."[40]

On the day of the protest, pro-life activists picketed the CBS-owned stations in Los Angeles, Chicago, Philadelphia, and St. Louis. Pressure was placed on other local stations as well, but only one affiliate—WXJT-TV in Greenville, Mississippi—refused to carry the show. Two stations—one in Omaha, the other in Cleveland—did agree to broadcast a half-hour response program entitled *A Matter of Choice*, prepared by pro-life groups. CBS Television City in Los Angeles received over 3,000 angry phone calls following the broadcast. The next day, leaders of a coalition of pro-life groups called for a boycott of CBS, the *Cagney & Lacey* series, and the fourteen sponsors whose commercials appeared in the controversial episode. None of these pressure efforts had any substantial impact.[41]

As for the ratings, *Cagney & Lacey* lost out not only to *An Early Frost*, which earned a 33 share, but also to ABC's *Monday Night Football*, which got a 32. *Cagney & Lacey* captured a little less than a fourth of the viewing audience that night, still not a bad number, given the heavy competition.

CHAPTER ELEVEN

From Ferment to Feedback

In the spring of 1986, NBC held a private conference in Tarpon Springs, a scenic resort town off the coast of Florida. The guests for this all-expenses-paid event—entitled "The Public Interest and an Interested Public"—had been carefully chosen. Leaders of over 20 organizations were there, including the National Gay and Lesbian Task Force, the PTA, the Southern Baptist Convention, and the American-Arab Anti-Discrimination Committee. NBC had organized five of these meetings since 1979. The agenda for this one was to "further dialogue and provide a sounding board for feelings—whether positive or negative—about the role of television in American life." During two intense days, participants heard lectures from academics and network executives on the social impact of media, the role of stereotypes, and "the whys and hows of TV programming." In a role-reversal session, advocacy group leaders were given the chance to "play standards and practices editors" by reviewing and discussing clips from sensitive programs. At the end of the meeting, conferees were invited to "develop a relationship with NBC," and encouraged to "sing out on the phone, let us know your feelings."[1]

This cleverly orchestrated event was emblematic of the institutionalized relationship which the networks had developed with advocacy groups. By the late eighties, skillfully fashioned industry strategies had transformed advocacy groups from a disruptive force into what network executives referred to as a "feedback system."

Societal changes had also played a role in the evolution of

this relationship. As the turbulence of the sixties and seventies waned, the number of activist groups declined, and pressures on many institutions subsided. But the most important stabilizing factor was the television industry's creation of self-protective mechanisms for circumscribing pressure from advocacy groups. These mechnaisms were developed because of network television's inherent vulnerability as a government-regulated, advertiser-supported medium. Network standards and practices departments took primary responsibility for creating and implementing these policies, but advertisers, affiliated stations, and producers also created their own protective mechanisms when they became the targets of outside pressure. Instituted during periods of heavy pressure, many of the mechanisms became permanent components in the complex machinery of network television.

Advocacy groups did have some success in their efforts to influence prime-time programming. The most effective groups were those whose goals were compatible with the network TV system, and whose strategies were fashioned with a keen awareness of how that system functioned. But in many cases, the success of these groups was influenced by external factors outside their control. The prominence and influence of these groups waxed and waned over time, corresponding to changes in the political climate or shifts in television industry practices and priorities. Gay activists were more influential in network TV when prime-time television, and the society at large, were particularly open to more liberal social mores; they were less successful in the early eighties during a period of conservatism in American politics. Groups concerned about alcohol and drug abuse found the television industry most receptive to their suggestions when that issue became important as a major public health concern. And women's groups were able to encourage some programs to feature more positive and prominent roles for women, as societal attitudes shifted and advertisers in turn sought to reach upscale female target audiences.[2]

For ethnic groups, the experience with the networks was mixed. Through intense pressure campaigns, and the use of government regulation as leverage, these groups were able to gain access to network television in the sixties and seventies.

But, like other groups, they succeeded in influencing program content only to the degree that network television was willing and able to accommodate them. Continued involvement with entertainment television required adapting to network policies, which effectively channeled the groups into more manageable relationships with the television industry. In time, militant groups were replaced by more moderate organizations. Many members of these newer groups were employees of the television industry, which put them in a very difficult position for adversarial efforts. Although they maintained amicable relations with the networks, they appeared to have little impact on prime-time programming, where ethnic characters and images remained marginalized.[3]

Those groups that tried to force broad changes in programming content—especially changes that ran counter to prevailing industry trends—were the least effective. Though highly publicized, the campaigns to reduce sex and violence in prime time enjoyed only brief success. They were able to modify program content for relatively short periods of time, but soon after the campaigns ended, sex and violence rose to previous levels.

While individual groups had varying degrees of success at influencing program content, the aggregate impact of advocacy groups on prime time has been substantial. Entertainment programming's function as an "environment" for commercial messages has played a fundamental role in this process. Advertisers sought certain kinds of programming to attract desirable audiences. This drive for ratings pushed entertainment television into provocative and controversial areas, thus drawing public attention and political pressure. To maximize gain and minimize risk the networks devised special content policies for handling sensitive issues. These policies served as part of the network filtering mechanism, selecting and shaping issues and images. In time, the policies evolved into standard narrative conventions, which were internalized by all segments of the television industry.

As a consequence, distinct patterns can be seen in the programming of entertainment television. While permeating the landscape of prime time, political and social issues are shaped to conform to the institutional demands of network television.

Important societal conflicts are extracted from the public sphere and injected into entertainment programs, where they are reduced to problems for individual characters. Political ideas are expressed as personal opinions. Controversial issues are consistently and carefully balanced within each program so that one clear argument cannot be discerned. Such purposefully ambiguous treatment has transformed prime-time programming into a psychological ink-blot test, open to varying interpretations.

As the networks continue to move toward stability in their relations with advocacy groups, several recent developments have begun to alter the relationship between the TV industry and the public: government deregulation, the rapid spread of cable and other new communications technologies, and shifts in media ownership patterns.

The efforts by broadcasters to undercut the legal powers of advocacy groups ultimately paid off. By the late seventies, major industries such as banking, trucking, and airlines convinced lawmakers that the economic advantages of deregulation were in the public interest. The broadcasting industry took advantage of this new political climate to push its own deregulatory agenda. The most sweeping deregulation took place in the first few years of the Reagan administration. FCC chairman Mark Fowler—a Reagan appointee—launched an aggressive program to "unregulate" broadcast television, quickly reversing many of the gains public interest groups had made. TV stations could now renew their licenses merely by sending in a postcard, and were no longer obligated to meet with community groups or to broadcast news and public affairs programs. As a result, the ability of public interest groups to challenge broadcasters was greatly diminished.[4]

Media reform groups struggled desperately to hold back the deregulatory tide. The organizations that had brought about the reforms of the seventies now found themselves on the defensive, fighting losing battles with the broadcast industry. And with significant reductions in foundation support, many of them had trouble just staying afloat.[5]

Long targeted for elimination by the industry, the Fairness

Doctrine became the focus of intense battles. Broadcasters argued that the Fairness Doctrine infringed upon their First Amendment rights. For years, the doctrine had been an annoyance to station managers because it made them vulnerable to criticism and required them to take protests seriously, sometimes obligating stations to provide free time to outside organizations. The battle to retain the doctrine was fought by an eclectic group of organizations with clearly divergent political positions. Among the policy's staunch supporters were: the ACLU, the right-wing Eagle Forum, consumer advocate Ralph Nader, and two former chairmen of the FCC—Newton Minow and Charles Ferris. The controversy came to a head in 1987. Broadcasters challenged the constitutionality of the Fairness Doctrine in the courts. Congress passed a bill to make the doctrine a clear part of the Communications Act, legislation which President Reagan promptly vetoed. Finally, the FCC, claiming the doctrine "no longer served the public interest," voted to eliminate the policy in August of 1987. Media reform groups immediately challenged that action, appealing the decision in the courts and vowing to get the Fairness Doctrine back on the books through Congress.[6]

The main justification for deregulation was the claim that the public interest would be better served if marketplace forces were allowed to operate, unfettered by government rules. Any limitations in the present system would be overcome by emerging new communication technologies which would create a more equitable and diverse system: satellite-delivered cable networks would cater to almost every special interest; new low-power broadcast stations would make it possible for virtually anyone to become a broadcaster; and public access cable channels would become the new "electronic soap box," providing a forum for those shut out by traditional broadcasting.

But by the end of the 1980s utopian promises of diversity and access had not been realized. Cable did create some new programming services that differed from their broadcast predecessors. However, the same economic forces that controlled American broadcasting from its beginnings shaped the development of cable. Most of the new special-interest cable channels were designed as advertising vehicles. The few programming services developed for blacks and Hispanics featured light entertainment fare, with a limited amount of news and public affairs programming. Shopping channels, music video channels, and travel channels were basically wall-to-wall advertising. Other cable networks filled their schedules with broadcast network reruns and theatrical films, offering viewers recycled images from the TV and movie past. Home video quickly established itself as a successful new industry by recycling these same images. Low-power television emerged as a tiny second-class service, with limited reach and little economic viability. Public access channels—with only feeble support from the government, and virtually none from the cable industry—remained an extremely marginalized service.[7]

Ownership within the cable industry became further concentrated. A handful of major corporations steadily bought up cable systems around the country, controlling more than half the national market by the late eighties. Vertical integration—where the same company owns and controls the means of program production as well as distribution—became the norm. Hence, as cable channels multiplied, the number of cable companies rapidly declined.[8]

In the mid 1980s the broadcast networks underwent unprecedented changes. The growth of cable television, the explosion of the home video industry, and the expansion of independent TV stations significantly reduced the networks' share of the audience. Takeover challenges or ownership changes occurred at all three networks: ABC was purchased by Capital Cities Communication, and General Electric bought NBC's parent company, RCA. Cable entrepreneur Ted Turner and a right-wing political group both tried unsuccessfully to buy out CBS, forcing that network to take drastic financial measures to fend off a takeover. There were intense pressures at all three networks to adhere to the bottom line. Cuts were made in a number of departments.[9]

Among the casualties of these cutbacks were the network standards and practices departments, whose staffs were reduced by 25 percent between 1986 and 1987. The networks could afford to reduce the size of these departments because deregulation and stabilized relations with advocacy groups had lowered the level of threat from outside forces. In the absence of pressure, mechanisms of content control could be relaxed. More important, the formulas and conventions for handling sensitive material were so well internalized by all participants in the creation of network entertainment programming that the system could continue with little intervention from the standards and practices editors. CBS discontinued routine involvement of standards and practices with certain prime-time series and assigned such responsibilities to the programming department. "The producers know the things that will raise a red flag and they will send them to us for consultation," explained George Dessart, CBS's vice president of program practices.[10]

A new era has arrived for the television industry. Network television has lost its monopoly on the audience, and the government's mechanisms for protecting the public interest have been dismantled.

Organized groups will continue to appeal to the electronic storytellers of our culture, urging them to incorporate messages into their plots, enhance the representation of certain groups, or eliminate content considered offensive or dangerous.

Though their share of the nationwide audience is on the decline, the three major networks may continue to hold a central place in American culture for some time to come, remaining the targets of various advocacy groups. However, emasculated government regulatory policies may make it increasingly difficult for such groups to have an impact on the networks. The networks may remain willing to "dialogue" with advocacy groups, as part of their continuing strategy for deflecting pressure. But without effective political leverage, groups are unlikely to have their demands and concerns taken seriously.

The policy of balance most likely will remain, whether or not efforts to reinstate the Fairness Doctrine succeed. Like the other protective mechanisms within the network TV industry, balance has become a firmly entrenched practice, a product of economic and institutional forces as well as regulatory policy. Balance is too valuable to be abandoned. Tightly woven into the operation of network television, it deflects criticism and protects advertisers and affiliates.

However, balance does nothing to facilitate access for advocacy groups. Its function is a pre-emptive one, designed to ward off action by organizations or individuals. Without the Fairness Doctrine as a weapon, advocacy groups will be left with virtually no avenues for legal recourse.

Any groups trying to influence the content of prime-time television will need to adjust their strategies to changes taking place within the industry. New groups will appear on the scene bearing new agendas and armed with new weapons. Some may be militant in their strategies, attempting to attack vulnerable parts of the television industry. A few groups may try to use strategies that were successful in network television to influence cable television. These efforts are unlikely to be effective in an industry which lacks a tradition of interaction with the public. Since the cable industry is not licensed by the government and has no legal requirement to serve the public interest, cable operators and cable networks have no obligation to listen to the complaints or concerns of outside groups.

Facing less-responsive broadcast networks and an uninterested cable industry, more groups are likely to appeal directly to the creators of entertainment programming. Such a trend is already apparent. The last few years have seen a rise in the number of organizations focusing their attention on the Hollywood creative community. Public health organizations and environmental groups, for example, have recently begun to approach producers and writers about how entertainment programming could play a role in educating the public about their issues. As long as such appeals are made in modes acceptable to the entertainment industry, they will be tolerated, and may even be encouraged. Such requests will be accommodated to the extent that they are compatible with the needs of entertainment television.

Given the industry's constant need for programming ideas, and its reliance on "headline" stories to draw large audiences, political and social issues will continue to provide a steady stream of material for prime-time television. With the drastic reduction in network documentaries, the disappearance of local public affairs programs, and the increase in the number of docudramas, more and more of the critical issues of the day will be packaged as entertainment, and shaped according to formulas that have been developed and refined to maximize audiences and profits.

It is clear that economic forces will continue to dominate American television. Any diversity created by the system will be market-driven. Variations in program offerings will be based on the needs of advertisers. But this is very different from the political and cultural diversity essential to a healthy democracy. The economic marketplace is not synonymous with the marketplace of ideas. While certain segments of the population may be well served as consumers, the public as a whole will not be well served as citizens. If present trends continue, voices in our pluralistic society will be further marginalized, circumscribed, or excluded.[11]

Through their efforts during the last several decades, advocacy groups have succeeded in raising critical issues about the accessibility and accountability of the nation's mass media. But both individually and collectively these groups have lacked the power and resources to truly open the system to a range of ideas. The public interest groups organized in the sixties to reform the broadcasting industry were gradually replaced by organizations with more limited goals. These groups adapted to the system instead of trying to change it. As electronic media become increasingly important in our lives, those with the power to tell the stories of our time will help shape how we see our world. If we are to have a communications system that fully serves the needs of American democracy, media policy will need to be placed high on the public agenda.

Notes

Chapter 1. Prime Time as Political Territory

1. Editorial, "Why Is ABC Doing Yuri Andropov's Job?," *New York Post*, Sept. 23, 1983. Interview with Josh Baran, Berkeley, Feb. 18, 1984. Interview with Herbert Gunther, Public Media Center, San Francisco, Feb. 17, 1984. Glenn Collins, "The Impact on Children of 'The Day After,' " *New York Times*, Nov. 7, 1983.
2. Rick Ruiz, "Rev. Falwell Organizing Boycott of Controversial Show's Advertisers," *Los Angeles Herald Examiner*, Nov. 16, 1983. "*The Day After:* A Gamble That Paid Off!," J. Walter Thompson U.S.A., Inc., Dec. 1, 1983. Interview with Reed Irvine, Accuracy in Media, Washington, D.C., March 5, 1986. At a ABC stockholders' meeting a few months after the broadcast of *The Day After*, Irvine introduced a resolution "calling for an investigation into whether ABC was being used as a channel for Soviet disinformation." Irvine also took some credit for the network's decision to produce *Amerika*, a 1987 miniseries about a hypothetical Soviet takeover of the U.S. See Jeff Gottlieb, "Prime-Time Patriotism: ABC Atones for *The Day After*," *Mother Jones* (Jan. 1987).
3. Jay Sharbutt, "Nuke Film Gets High Ratings," *Los Angeles Times*, Nov. 22, 1983. "The Nightmare Comes Home," *Time*, Oct. 24, 1983. Tom Shales, "Must Viewing for the Nation," *Washington Post*, Nov. 18, 1983.
4. For a discussion of pressures on the film industry, see Garth Jowett, *Film: The Democratic Art* (Boston: Little, Brown, 1976). Todd Gitlin used the term "contested zone" to describe the state of television during the turbulent period of the 1970s. Todd Gitlin, *Inside Prime Time* (New York: Pantheon, 1983). For a transcript of the 1981 industry symposium, see Lee Margulies, ed., "The Prolifer-

ation of Pressure Groups in Prime Time Symposium," *Emmy Magazine*, Summer 1981.

5. Harry Waters, "Life According to TV," *Newsweek*, Dec. 6, 1982.
6. George Gerbner and Larry Gross, "Living with Television: The Violence Profile," *Journal of Communication* (Spring 1976). Todd Gitlin, *Inside Prime Time* (New York: Pantheon, 1983), 333–34.
7. Vincent Mosco, "The Invisible Majority," *Communication Perspectives* (July 1981). Leonard J. Theberge, ed., "Crooks, Clowns, and Conmen: Businessmen in TV Entertainment," the Media Institute, Washington, D.C., 1981. Les Brown, "Study Finds Stereotyping in TV Casts," *New York Times*, Oct. 30, 1979. "Gaps Found between TV Families and Reality," *Broadcasting*, Sept. 2, 1985.
8. Quote from William W. Wipinsinger, president, International Association of Machinists, in "Union Media Monitoring Project," n.d., p. 2 (courtesy of International Association of Machinists, Washington, D.C.).
9. Howard Rosenberg, "Pressure Groups and TV—The Shadow over Ojai," *Los Angeles Times*, May 13, 1981.
10. Muriel Cantor, "The Politics of Popular Drama," *Communication Research* (Oct. 1979), 387–406.
11. Carrie Rickey, "Why They Fight; Subjects' Rights and the First Amendment," *American Film* (Oct. 1981). Jack Slater, "Gerber Hits TV Pressure Groups," *Los Angeles Times*, Sept. 25, 1980. See Kathryn Montgomery, "Gay Activists and the Networks," *Journal of Communication* (Summer 1981).
12. Richard Levinson and William Link, *Stay Tuned: An Inside Look at the Making of Prime-Time Television* (New York: St. Martin's, 1981), 191.

Chapter 2. Television Under Siege

1. NCCB's 175 members included luminaries such as John Kenneth Galbraith, Archibald MacLeish, Bill Cosby, Harry Belafonte, Henry Morgan, Mike Nichols, and Leslie Uggams. Tom Macken, "Networks Under Fire," *Newark Evening News*, Jan. 9, 1969.
2. J. Frank Reel, *The Networks: How They Stole the Show* (New York: Charles Scribner's Sons, 1979), 67.
3. See Erik Barnouw, *The Sponsor* (New York: Oxford Univ. Press, 1978).
4. Explained one advertiser: "If we don't screen out controversial people, we will be hurting the sales of the product we are trying

to sell. Therefore, not to screen would be unbusinesslike and violate the trust of stockholders." John Cogley, *Report on Blacklisting* (New York: Fund for the Republic, 1956), 100.

5. Barnouw, *The Sponsor,* 49. See also Karen Sue Foley, *The Political Blacklist in the Broadcast Industry: The Decade of the 1950's* (New York: Arno Press, 1979).
6. Arnold Shankman, "Black Pride and Protest: The *Amos 'n' Andy* Crusade," *Journal of Popular Culture* (Fall 1978), 248–49. Cheryl Chisholm, "The *Amos 'n' Andy* Controversy" (unpublished paper, UCLA, 1983).
7. The NAACP had been working in Hollywood to improve the image and employment of blacks in the film industry. Its leaders believed they had made some gains in motion pictures. Thomas Cripps, "*Amos 'n' Andy* and the Debate over American Racial Integration," in John E. O'Connor, ed., *American History/American Television: Interpreting the Video Past* (New York: Frederick Ungar, 1983). "Negro Performers Win Better Roles in TV Than in Any Other Entertainment Medium," *Ebony* (June 1950).
8. Cripps, "*Amos 'n' Andy,*" 38. J. Fred MacDonald, *Blacks and White TV: Afro-American in Television since 1948* (Chicago: Nelson-Hall, 1983), 38–50.
9. Stockton Hellfrich, "Self-Regulation by Networks," *Journal of Broadcasting* (Spring 1957).
10. Cobbett S. Steinberg, *TV Facts,* 1980, Facts on File, Inc., p. 142. James L. Baughman, "Television in the 'Golden Age': An Entrepreneurial Experiment," *The Historian* (Feb. 1985), 178–79.
11. "Sniping at Radio-TV: A National Pasttime," *Broadcasting,* April 19, 1956. "Quiz Scandal Spotlight May Alter Basic Structure of TV," *Advertising Age,* Nov. 9, 1959. "Hollywood Drafts Production Code to Guide Film, Live TV Programming," *Broadcasting*Telecasting,* April 9, 1956.
12. As early as 1952, the House of Representatives held hearings to investigate "moral and otherwise offensive matter" in television programming. By 1954 the Senate had taken up the investigation, this time focusing on the possible contribution that television programming made to crime and juvenile delinquency. See Willard D. Rowland, Jr., *The Politics of TV Violence* (Beverly Hills: Sage, 1983), 99, 101–5.
13. Hellfrich, "Self-Regulation by Networks." Harvey C. Jassem, "An Examination of Self-Regulation of Broadcasting," paper presented to International Communication Association, Dallas, Texas, May 1983.

14. "Broadcasting's Year of Trouble," *Broadcasting,* Feb. 15, 1960. See Erik Barnouw, *Tube of Plenty: The Development of American Television.* (New York: Oxford Univ. Press, 1975). See also Kent Anderson, *Television Fraud: The History and Implications of the Quiz Show Scandals* (Westport, CT: Greenwood Press, 1978).
15. Barnouw, *The Sponsor.*
16. Muriel G. Cantor, *Prime Time Television: Content and Control* (Beverly Hills: Sage, 1980), 69.
17. "Now a Deep Scar in TV's Image," *Broadcasting,* Nov. 9, 1959. Christopher Sterling and John M. Kittross, *Stay Tuned: A Concise History of American Broadcasting* (Belmont, CA: Wadsworth, 1978), 572.
18. Cynthia Lowry, "TV Censor's Job: Listening to All Those Complaints," Associated Press column, Aug. 22, 1961 (courtesy of Television Information Office, New York City).
19. Val Adams, "Willy Loman Irks Fellow Salesmen," March 18, 1966 (courtesy of Television Information Office, New York City).
20. "Italian-Americans to Boycott 'Untouchables,' " *Broadcasting,* March 13, 1961. "Against Untouchables," *New York Times,* March 13, 1961. "They Want Full Surrender," *Broadcasting,* March 20, 1961.
21. Jack Gould, "Disturbing Pact: Compromise on 'Untouchables' Holds Dangers for Well-Being of TV," *New York Times,* March 26, 1961. "Power Plays Topple Fall Lineup," *Broadcasting,* March 20, 1961.
22. Gould, "Disturbing Pact."
23. See MacDonald, *Blacks and White TV.*
24. Peter Bart, "NAACP Charges Film and TV Bias," *New York Times,* July 12, 1966. Harry Bernstein, "Negroes in Hollywood: Opportunity at Last," *Journal American,* March 20, 1966. Richard Gehman, "Black and White Television?," *TV Guide,* June 27, 1964. "Rights Movement Pointing for the Networks," *Insider's Newsletter,* Jan. 4, 1965.
25. "ABC Steps Up Integration on Daytime Soapers," *Variety,* April 14, 1965. Val Adams, "Georgia Stations Snub 'I Spy,' " *The Atlanta Constitution,* Sept. 13, 1965. Untitled newspaper article, Nov. 8, 1965 (courtesy of Television Information Office).
26. MacDonald, *Blacks and White TV,* 108.
27. Richard K. Doan, "Two Stations Accused of Bias," *New York Herald Tribune,* April 14, 1964.
28. Fred M. Friendly, *The Good Guys, the Bad Guys, and the First Amendment* (New York: Random House, 1976), 89–102.
29. Fred P. Graham, "Court Upholds Plea by Negroes for Voice in TV License Renewal," *New York Times,* March 26, 1966.

30. "The Struggle over Broadcast Access," *Broadcasting*, Sept. 20, 1971.
31. J. A. Grundfest, *Citizen Participation in Broadcast Licensing Before the FCC* (Santa Monica, CA: Rand Corporation, 1976). Donald L. Guimery, *Citizens Groups and Broadcasting* (New York: Praeger, 1976). "Citizens with Clout," *TV Guide*, March 8, 1975.
32. Nicholas Johnson, *How to Talk Back to Your Television Set* (Boston: Little, Brown, 1970). "The Struggle over Broadcast Access," *Broadcasting*, Sept. 20, 1971.
33. FCC Members Trade Charges on License Renewal in Bias Case," *New York Times*, July 13, 1968. Barry Cole and Mal Oettinger, *Reluctant Regulators: The FCC and the Broadcast Audience* (Reading, MA: Addison-Wesley, 1978).
34. By 1977, there were more than 400 media reform groups listed in NCCB's Citizen's Media Directory. Cherie Lewis, *Television License Renewal Challenges by Women's Groups* (unpublished doctoral dissertation, University of Minnesota, 1984). See Cole and Oettinger, *Reluctant Regulators.* See also Willard D. Rowland, Jr., "The Illusion of Fulfillment: The Broadcast Reform Movement," *Journalism Quarterly* (Dec. 1982).
35. National Organization for Women Position Paper read at NCCB Conference, Oct. 26, 1976. Lewis, *Television License Renewal Challenges by Women's Group.*
36. Paul Hoffman, "German TV Roles Assailed as Slur," *New York Times*, July 15, 1968. "Ethnic Groups Seek Ban on Slurs on TV," *New York Times*, July 24, 1968. "Custer Series Still On," *New York Times*, July 24, 1967.

Chapter 3. And Then Came *Maude*

1. *Maude* episode, "Maude's Dilemma," broadcast Nov. 14, 1972, CBS.
2. Aljean Harmetz, "Maude Didn't Leave 'Em All Laughing," *New York Times*, Dec. 10, 1972.
3. Todd Gitlin, *Inside Prime Time* (New York: Pantheon, 1983), 203–20. Joseph Turow, "Unconventional Programs on Commercial Television: An Organizational Perspective," in James S. Ettema and D. Charles Whitney, eds., *Individuals in Mass Media Organizations: Creativity and Constraint* (Beverly Hills: Sage, 1982). Horace Newcomb and Robert S. Alley, *The Producer's Medium* (New York: Oxford Univ. Press, 1983), 176. Tim Brooks and Earle Marsh, *The Complete Directory for Prime Time Shows, 1946–Present* (New York: Ballantine, 1979), 19–22.

4. Joseph Morgenstern, "Can Bigotry Be Funny?," *Newsweek*, Nov. 29, 1971. A number of social critics took issue with this assumption, and in at least one scholarly study, the results showed that the producers' intent may have backfired. Those viewers who agreed with Archie Bunker's conservative point of view in the first place tended to have their opinions reinforced by the show, rather than challenged. See Neil Vidmar and Milton Rokeach, "Archie Bunker's Bigotry: A Study in Selective Perception and Exposure," *Journal of Communication* (Winter 1974), 36–47. As the authors of this article concluded: ". . . the program encourages bigots to excuse and rationalize their own prejudices. . . . Already there is evidence that impressionable white children have picked up, and are using, many of the racial slurs which Archie has resurrected, popularized, and made 'acceptable' all over again."
5. "The Team Behind Archie Bunker & Co.," *Time*, Sept. 25, 1972.
6. Ibid.
7. Robert B. Beusse and Russell Shaw, "Maude's Abortion: Spontaneous or Induced?," *America* (Nov. 1973). The preceding from a Population Institute pamphlet, quoted in the article. Interview with Norman Fleishman, Los Angeles, Nov. 13, 1984. Telephone interview with David Poindexter, Aug. 28, 1985.
8. Interview with Fleishman. Interview with Poindexter.
9. Beusse and Shaw, "Maude's Abortion." Harmetz, "Maude Didn't Leave 'Em All Laughing." In a letter to the editor of *America* (written in response to the Beusse and Shaw article), Mrs. Theodore O. Wedel, board member of the Population Institute, explained the way the idea came about for the episodes: "The original interest in the population issue on the part of the producers of *Maude* was stimulated by an address delivered by a physicist from the California Institute of Technology. In the audience was the producer of *Maude*." "The 'Maude' Case: Pressure or Persuasion?," *America*, Dec. 15, 1973. When I asked Norman Lear about this, the producer was vague in his recollection of exactly how the idea came about. Interview with Norman Lear, Los Angeles, Nov. 11, 1985.
10. Harmetz, "Maude Didn't Leave 'Em All Laughing."
11. See Muriel Cantor, *Prime Time TV: Content and Control* (Beverly Hills: Sage, 1980). The operation of the standards and practices departments will be described in detail in Chapters 4 and 10. Gitlin, *Inside Prime Time*. Turow, "Unconventional Programs on Commercial Television." Newcomb and Alley, *The Producer's Medium*, 176.

12. Interview with Barbara Brogliatti, Tandem Productions, Los Angeles, Aug. 30, 1978. Interview with Virginia Carter, Los Angeles, Nov. 27, 1984.
13. Interview with Lear.
14. Ibid. Despite the conscious attempt to balance the story by adding this character, the way the whole scene was presented undermined its balancing effect. The woman asks Maude to watch her children in the car for a few minutes. Though we never see the children (they are off-screen), Maude is shown screaming hysterically at them because they are obviously out of control and causing great havoc. *Maude* episode, "Walter's Dilemma," broadcast Nov. 21, 1972, CBS.
15. Memo from Norman Lear to Paul King, Aug. 20, 1972 (courtesy of Embassy Television).
16. Harmetz, "Maude Didn't Leave 'Em All Laughing." Interview with Lear.
17. Interview with Lear. Interview with Poindexter.
18. Letter from David Poindexter to Norman Lear, Dec. 4, 1972. "NOW Suit over Maude Abortion Shows Denied," *Los Angeles Times,* Nov. 25, 1972 (courtesy of Embassy Television). The Peoria station refused only to air the second episode.
19. Estimates of the numbers of letters varies. In one network memo, reports from different departments calculating the ratio of con versus pro letters ranged from 2:1 to 45:1. According to Richard Levinson and William Link, CBS received 24,000 letters about the show. Richard Levinson and William Link, *Stay Tuned:* An Inside Look at the Making of Prime-Time Television (New York: St. Martin's, 1981). Cecil Smith, "Maude's Abortion Evokes Protests," *Los Angeles Times,* Nov. 29, 1972.
20. Letter from Monsignor Eugene V. Clark, unsigned, to Richard W. Jencks, president, CBS/Broadcast Group, Nov. 21, 1972. Memo to Norman Lear, Dec. 1, 1972 (courtesy of Embassy Television). According to an editorial in *America,* Dec. 15, 1973, the protesters were given a meeting with top executives at CBS.
21. See Fred W. Friendly, *The Good Guys, the Bad Guys, and the First Amendment: Free Speech vs. Fairness in Broadcasting* (New York: Random House, 1976). Steven J. Simmons, *The Fairness Doctrine and the Media* (Berkeley: Univ. of California Press, 1978), 152–3. Cherie Lewis, *Television License Renewal: Challenges by Women's Groups* (unpublished doctoral dissertation, Univ. of Minnesota, 1984), 95.
22. Re Complaint by Diocesan Union of Holy Name Societies of Rock-

ville Centre and Long Island Coalition for Life Concerning Fairness Complaint, Re Station WCBS-TV, New York, NY, June 12, 1973.

23. Ibid. Richard A. Blake, "O, Maude, Poor Maude," *America*, Sept. 1, 1973.
24. "Critics Assail Pro-Abortion View," *The Tidings*, Dec. 15, 1972. Attachment to letter dated Nov. 27, 1972, from Gene Denari, KBAK-TV to Tom Downer, Program Practices, CBS.
25. Planned Parenthood—World Population, Staff Memorandum, Aug. 16, 1973.
26. Interview with Fleishman.
27. Letter, "Stop Immorality in Media," dated May 11, 1973 (courtesy of Embassy Television).
28. "The Team Behind Archie Bunker & Co.," *Time*, Sept. 25, 1972. " 'Bridget Loves Bernie' Attacked by Jewish Groups," *New York Times*, Feb. 7, 1973. "Some Jews Are Mad at Bernie," *New York Times*, Feb. 11, 1973. Albert Krebs, " 'Bridget Loves Bernie' Dropped from C.B.S. Schedule for Fall," *New York Times*, March 30, 1973.
29. Robert G. Pekurny and Leonard D. Bart, " 'Sticks and Bones': A Survey of Network Affiliate Decision Making," *Journal of Broadcasting* (Fall 1975).
30. Interview with Poindexter.
31. See Simmons, *The Fairness Doctrine and the Media*, 153.
32. Interview with Lear.
33. Norman Lear himself was a founding member of the ACLU Foundation Board and served as its executive director from 1970 to 1984.
34. "Protest on *Maude* Widens," New York *Daily News*, Aug. 8, 1973. " 'Maude' Sponsorship Decline Laid to Abortion Foes" *New York Times*, Aug. 10, 1973. New York *Daily News*, Aug. 6, 1973. "Protests Mount over 'Maude' Rerun," *The Tidings*, Aug. 10, 1973.
35. "Hub CBS Affil Will Run 'Sticks' But Nixes 'Maude' Abortion Segs," *Daily Variety*, Aug. 2, 1973.
36. "The Way It Should Work," *Broadcasting*, Aug. 27, 1973.
37. " 'Maude' Sponsorship Decline Laid to Abortion Foes," *New York Times*, Aug. 10, 1973. Most likely the protesting groups had not known who the specific sponsors of the show would be for the two reruns, because that information is not made available by the networks. They did know, however, who advertised regularly in those time slots, as well as who had advertised in the previously run abortion episodes. Certainly those advertisers who had been targeted after the first broadcast would be reluctant to set themselves up for a potential boycott again. "Maude Is Target of Abor-

tion Forces," Aug. 21, 1973 (courtesy of Embassy Television). Ironically, Henry Hayes, the media director for Pepsi-Cola, had just written his guest editorial in the trade publication *Media Decisions*. As the *New York Times* noted, "the editorial was written and on the presses before Pepsi withdrew its advertising from the two reruns of the 'Maude' series that dealt with abortion. The magazine did have time, however, to get a few paragraphs up front explaining the timing." "Advertising: A Magazine Pitch," *New York Times*, Sept. 19, 1973.

38. As American Home Products later explained in a letter to a viewer: "The withholding of commercials from the two-part Maude re-run episode was based on the large number of station affiliate defections which left us with incomplete coverage for our products, and was not based on program content."
39. "Maude's Abortion Stirs New Protests," *Los Angeles Herald Examiner*, Aug. 16, 1973. Internal memorandum, Tandem Productions, Aug. 1973 (courtesy of Embassy Television).
40. Press release, National Association for the Repeal of Abortion Laws, New York, Aug. 17, 1973.
41. Press release, American Civil Liberties Union, New York, Aug. 30, 1973.
42. "Maude, 'Sticks' Action and Reaction," *Daily Variety*, Aug 22, 1973.
43. "Program Pressure," *New York Times*, Aug. 24, 1973.
44. Beusse and Shaw, "Maude's Abortion." The Population Institute answered these allegations in a follow-up article, Dec. 15, 1973, in which the organization's leaders argued that they had never pressured the show's creators. "The 'Maude' Case: Pressure or Persuasion?," *America*, Dec. 15, 1973.
45. "Advertising: A Magazine Pitch," *New York Times*, Sept. 19, 1973.
46. "Lear: 'Maude' Is More Than a Ruined Roast," *The Evening Bulletin*, Aug. 17, 1973.

Chapter 4. Managing Advocacy Groups

1. "Chicanos' question: What About Us?," *Broadcasting*, June 28, 1971.
2. By the 1970s all three networks had standards and practices departments. They were structured and functioned quite similarly, although they had different names at each of the networks. At NBC, the name was Broadcast Standards; at CBS it was Program Practices; and at ABC it was Broadcast Standards and Practices. A parallel function was served by the network programming depart-

ments, whose staff also supervised the development of new programming and the writing and production of existing programs. In their production supervisory capacities, both departments operated similarly, assigning staff to serve as liaison with outside producers. In contrast to standards and practices departments, programming departments were responsible for ensuring ratings success. Hence, they operated under a different set of imperatives and constraints from those of standards and practices. See Muriel Cantor, *Prime Time TV: Content and Control* (Beverly Hills: Sage, 1980). See also Robert Pekurny, *Broadcast Self-Regulation: A Participant-Observation Study of the National Broadcasting Company's Broadcast Standards Department* (unpublished doctoral dissertation, Univ. of Minnesota, 1977). Joseph Turow uses the term "boundary personnel," a concept borrowed from organizational theory, to characterize the role of standards and practices executives in dealing with outsiders. Joseph Turow, "The Influence of Pressure Groups on TV Entertainment: A Framework for Analysis," in Bruce Watkins and Willard D. Rowland, Jr., eds., *Interpreting Television: Current Research & Perspectives* (Beverly Hills: Sage, 1984).

3. "TV Prime Target for Pressure Groups," *Broadcasting,* April 10, 1961.
4. The term "manageable" is my own label for describing the way the industry has handled advocacy groups. It was inspired by a remark that Grant Tinker, who later became president of NBC, made at the industry-sponsored 1981 "Proliferation of Pressure Groups" symposium, when he referred to most advocacy groups involved with TV as "manageable." See Lee Margulies, ed., "Proliferation of Pressure Groups in Prime Time Symposium," *Emmy Magazine,* Summer 1981.
5. See Francisco Lewels, *The Uses of the Media by the Chicano Movement* (New York: Praeger, 1974). *New York Post,* April 4, 1970.
6. "Mexican-Americans Seek New Film, TV Image," *Los Angeles Herald-Examiner,* Oct. 10, 1970. Telephone interview with Ray Andrade, July 25, 1985.
7. Interview with Andrade.
8. See Barry Cole and Mal Oettinger, *Reluctant Regulators: The FCC and the Broadcast Audience* (Reading, MA: Addison-Wesley, 1978), 203–41.
9. Interview with Tom Kersey, Los Angeles, Dec. 20, 1984.
10. Ibid. According to Ray Andrade, Justicia never got the $10 million demanded from ABC, though the network did offer the group some concessions at the local level. ABC agreed to hire a Chicano

and to funnel additional money into local public affairs programs on Chicano issues.

11. Interview with Kersey.
12. See Michael Real, "*Marcus Welby* and the Medical Genre," in *Mass-Mediated Culture* (New York: Prentice-Hall, 1977). "Can Mr. Novak Keep Teachers Happy?," *Look,* Oct. 6, 1964. As the article had reported, the group paid particular attention to Mr. Novak's grammar: "Since he is supposed to be an English teacher, [Novak's] grammar in classroom scenes must be perfect. If it isn't, a flood of erudite mail descends on the producer pointing out his error. In its report at the end of the first season's filming, the panel gravely regretted that Mr. Novak still slipped into such common mistakes as saying 'like' instead of 'as' and 'can' for 'may.' As for slang, the panel will occasionally let him use an 'okay' outside the classroom, but shudders if he refers to his students as 'Kids.' "
13. "Chicanos' Question: What About Us?" Interview with Andrade.
14. Interview with Kersey.
15. "Justicia Now Moves Against NBC-TV," *Broadcasting,* Aug. 2, 1971. Interview with Meta Rosenberg, Aug. 24, 1987.
16. Interview with Andrade. Interview with Jay Rodriguez, NBC, Burbank, Nov. 30, 1984. Tim Brooks and Earle Marsh, *The Complete Directory to Prime Time Network TV Shows, 1946–Present* (New York: Ballantine, 1979), 450. Later on in the series, one episode featured Ricardo Montalban as a mild-mannered Pancho Villa. As Meta Rosenberg described it: "Pancho Villa was a great Mexican general and he had a very bad toothache. He didn't trust any of the dentists in Mexican villages, so he came to this little town to have his tooth taken out. The residents heard the bandit was coming to town and thought Pancho Villa was going to attack them. Pancho Villa rides in and he's this mild man who wants to have his teeth pulled. It was really a positive image." Andrade says he might have taken a different position on the series if he'd known how the episode was going to be handled. "We thought the part was stereotyped," he explained. "What we didn't know was that it was going to be played by Ricardo Montalban. If we'd known that, we wouldn't have objected because he doesn't play stereotyped roles." Interview with Rosenberg.
17. Interview with Andrade.
18. Penny Anderson, "The Real Chico Behind Chico." *New York Times Syndicate,* 1975; telephone interview with James Komack, Aug. 11, 1987; interview with Rodriguez. According to an interview that

Andrade did with the *Herald Examiner,* "The network people . . . felt that if I was connected with the show, they wouldn't have any hassles." Frank Torrez, "Man in the Middle," *Herald Examiner,* n.d. (courtesy of Ray Andrade).

19. Telephone invterview with Felix Gutierrez, Sept. 10, 1986. Interview with Andrade. Dave Kaufman, " 'Chico' Associate Producer Andrade Unhappy over Show's Chicano Image," *Daily Variety* (courtesy of Ray Andrade).
20. Interview with Kersey.
21. From *Nosotros Casting Directory Guide,* 1976–77. John M. Wilson, "Optimistic Mood at Nosotros Event," *Los Angeles Times,* June 12, 1984. John Volano, "Nosotros Holds 15th Award Show," *Los Angeles Times,* June 15, 1985.
22. Interview with Tony Cortez, Los Angeles, July 1978. Presentation by Francesca Friday, Media for Impact Conference, University of Southern California, Feb. 1982. The short-lived ABC comedy series *Condo* is a case in point. *Condo* lasted one season—1982–83. The show revolved around an upwardly mobile Hispanic family who lived next door to an Anglo family. Like the earlier *Bridget Loves Bernie,* the series featured an interracial marriage between the children of the two families. This presented plenty of opportunity for ethnic-based humor. The fact that several of the cast were Nosotros members made it difficult for the group to take an official position to the show, even though they were somewhat unhappy with it when the network pre-screened the pilot before them. Tony Garcia, unpublished seminar paper, Univ. of California, Los Angeles, March 1983.
23. Though the label "one voice concept" was taken from a network executive, for the most part, the labels I have given to the elements of this "management" system are my own invention. I have, however, often relied on the language and terminology used by the network executives (and often by the advocacy groups as well) for my categories. The research for this chapter is based on many interviews with various network executives and advocacy group representatives over a number of years. Some of the practices differed slightly from one network to the other, but the general policies were virtually the same.
24. See Cantor, *Prime Time TV,* 63–96.
25. Interview with Bettye Hoffman, NBC, New York, April 13, 1981.
26. Fred Smedley, "This Gray Panther Lashes out at Media," *Intelliger Journal,* Lancaster, Pennsylvania, Nov. 17, 1978. Interview with Lydia Bragger, New York, April 15, 1981.

27. Interview with Bragger. Interview with Hoffman.
28. Interview with Hoffman.
29. Interview with Bragger. According to Todd Gitlin, one of the advertisers, General Electric, did pull out because the show wasn't "balanced" enough. Todd Gitlin, *Inside Prime Time* (New York: Pantheon, 1983), 256–57.
30. The three networks maintained somewhat different policies regarding the release of scripts to outsiders. CBS's official policy was not to pre-screen; both NBC and ABC were more likely to permit it. And ABC's Alfred Schneider defended the policy of making scripts available because it increased "knowledgeability on the network's part." Kersey indicated that programs were frequently pre-screened, although he insisted they would not "pre-screen on demand." Interview with Kersey.
31. Though some labeled the series as "the Indian version of *Roots*," Wolper denied referring to *Hanta Yo* that way. Howard Rosenberg, "Sioux Protest *Hanta Yo*," *Los Angeles Times*, April 16, 1980. Interview with Kersey. Dwight Whitney, "The Warpath Runneth Over," *TV Guide*, May 19, 1984.
32. Such attempts to use insiders as a hedge against criticism had been used in the past, but on a sporadic basis. Fifteen years or so before the real surge in pressure group activity, one of the network standards and pratices departments kept getting letters complaining about the protrayal of Native Americans in one of the network's series. The manager of the department answered such complaints with a standard letter, which pointed out that the network had consulted with one of its employees, who just happened to be Cherokee. Since the man had assured the editors that the TV series was not offensive to him, outside complaints really had no justification. The network repeatedly used this "in-house Indian" as its staff specialist on Native American affairs, even though this was not an official part of the man's job. Finally, the employee informed the standards and practices department that he no longer wanted to be used to endorse the network's portrayals of Native Americans.

    Only local stations were directly licensed by the FCC and required to file annual compliance reports related to Equal Employment Opportunity requirements. But the networks were carefully watched by the FCC during this period of time for their EOC performance, and the threat of more formal requirements from the federal government was always present. See Cole and Oettinger, *Reluctant Regulators*. See also U.S. Commission on Civil Rights,

Window Dressing on the Set: Women & Minorities in Television, 2 vols. (Washington, D.C.: Government Printing Office, 1977, 1979).

33. A few producers occasionally dealt directly with advocacy groups, usually for script consultation purposes, and often at the insistence of the networks. But none developed a systematic approach for dealing with outside organizations on a regular basis, nor did any other production company hire a staff person exclusively for this purpose. Interviews with Virginia Carter, Aug. 30, 1978, Nov. 27, 1984. Even with this system, Lear's comedies never dealt with abortion again. Charlie Hauck, who began producing the *Maude* series for Lear after the controversial abortion episodes, reported the following incident: "I did a show in which Maude's 18-year-old nephew comes to visit, and his girlfriend follows him. She's pregnant, there's a whole show about that, in which we wouldn't even mention abortion, considering all the options and everything, the word abortion never even came up. . . . I won an award from the Population Institute so whatever we did we must have talked about the responsibilities of parenthood but never talked about the possibilities of ending it." Interview with Charlie Hauck, Los Angeles, July 11, 1985.
34. In an episode of *All in the Family* involving an attempted rape of Edith Bunker, Carter coordinated an entire educational campaign around the show. She worked with a rape crisis center not only on the script but also on the development of study guides which provided discussion questions based on the program. These materials were mailed out to rape treatment centers, police departments, and various organizations, announcing the broadcast date. When the show aired, Carter remembers, "people gathered in living rooms or police stations or whatever and the rape crisis centers and watched the show." Interview with Carter, 1984.
35. Pluria Marshall wrote to Lear, urging him to reconsider his decision and threatening a boycott of sponsors if the producer refused to cooperate. "Black children, who watch TV an average of six hours a day," Marshall wrote, "desperately need positive black images. Writing the father out of *Good Times* signifies the death of prime-time TV's only positive black adult male character." Lear's response to the protest was that the *Good Times* decision was irrevocable. The producer pointed to the father in *The Jeffersons* and Fred Sanford of *Sanford and Son* as examples of positive male images for blacks in his shows. Dissatisfied with Lear's answer, Marshall went to New York and met with CBS executives, who assured him that the program would always have a strong black

male figure in the cast. No boycott was called, and Marshall seemed reasonably satisfied with CBS's promise. He later observed that "the program did in fact improve considerably thereafter." Interview with Pluria Marshall, Washington, D.C., June 1983. Letter from Pluria Marshall to Thomas Wynan, CBS, 1980 (courtesy of National Black Media Coalition). As J. Fred MacDonald has observed: "The comedic formula developed by Lear and Yorkin was a microcosm of that historic synthesis achieved during the 1970's with regard to blacks in TV. On the one hand, there was exposure of black talent—more roles, more employment, more black-centered programs than in the past. Yet, there was an almost total relegation of blacks to comedies." J. Fred MacDonald, *Blacks and White TV: Afro-Americans in Television since 1948* (Chicago: Nelson-Hall, 1983), 176.

36. License challenges by citizen groups were not the only reason broadcasters sought protective legislation. They were also alarmed by recent FCC decisions in "comparative renewal" cases. The proposed legislation would have barred the FCC from considering competing applications for broadcast licenses during renewal time. For a full discussion of this issue, see Irwin Krasnow, Lawrence D. Longley, and Herbert A. Terry, *The Politics of Broadcast Regulation* (New York: St. Martin's, 1982), 206–39. See also Willard J. Rowland, Jr., *The Politics of TV Violence* (Beverly Hills: Sage, 1983), 141–49. Untitled broadcast industry newsletter (courtesy of the Television Information Office).
37. Krasnow et al., *The Politics of Broadcast Regulation*, 206–39.
38. Les Brown, "Broadcasters at Convention Strike Back at Activist Critics," *New York Times*, March 19, 1974.

Chapter 5. Invisibility and Influence

1. Richard Levinson and William Link, *Stay Tuned: An Inside Look at the Making of Prime-Time Television* (New York, St. Martin's, 1981), 102–37.
2. Ibid.
3. Ibid., 132–33.
4. Interviews with Newton Deiter, Los Angeles, April 27, June 8, 1978, April 11, July 17, 1979.
5. Kathryn Montgomery, *Gay Activists and the Networks: A Case Study of Special Interest Pressure in Television* (unpublished doctoral dissertation, Univ. of California, Los Angeles, 1979). Richard M. Levine,

"How the Gay Lobby Has Changed Television," *TV Guide,* May 30, 1981. Richard M. Levine, "Our Only Allies Now Are Our Worst Enemies," *TV Guide,* June 6, 1981.

6. Interview with Ron Gold, New York, May 2, 1979.
7. Ibid.
8. Ibid. Interview with Loretta Lotman, Los Angeles, May 24, 1979.
9. Throughout this protest, gays maintained that their major concerns were with the stereotypical behavior that could serve to reinforce negative misconceptions about homosexuality. According to Loretta Lotman, they were also concerned about the possibly damaging effect such a program might have on gay civil rights legislation. They were particularly disturbed about the notion in the program that the molested boy had to be continually reassured that he was "still a man."
10. Interview with Lotman.
11. Ibid.
12. Press release, National Education Association, Washington, D.C., Oct. 4, 1974. For a full account of the campaign to change the classification of homosexuality in the American Psychiatric Association, see Ronald Bayer, *Homosexuality and American Psychiatry: The Politics of Diagnosis* (New York: Basic Books, 1981).
13. "Outrage and Outcry," *Washington Post,* Oct. 8, 1974.
14. "Mopery Episode on 'Welby' Cues Protests and Affil Defections," *Daily Variety,* Oct. 2, 1974. "Outrage and Outcry."
15. "Outrage and Outcry." Press release from WPVI-TV, Philadelphia, Sept. 25, 1974.
16. "TV: Welby Tackles Child Molestation," *New York Times,* Oct. 8, 1974. Several of the stations that carried the episode added disclaimers to their broadcasts, and at least one of them gave the gay activists "equal time." The ABC station in Washington, D.C., preceded its broadcast with an announcement asserting that the program was not about homosexuality: "WMAL-TV has received a number of objections to our broadcast of the following episode of ABC's *Marcus Welby.* These objections have been based on the sincere belief that the program reinforces a strongly held, widely prevalent, and misconceived stereotype of homosexuals. While great care has been taken in the writing of the script, and production of the program, it is conceivable that some viewers with existing prejudices could misconstrue the program and have their bias reinforced. For that reason, WMAL-TV wishes to state that this program is intended to deal with the subject of child molestation, and not homosexuality, and further that available evidence indi-

cates that homosexuals are no more likely to be involved in child molestation than heterosexuals." WMAL transcript, Oct. 8, 1974.

17. Interview with Gold. Interview with Lotman. "Gays Still Protesting *Welby*," *Washington Post*, Oct. 10, 1974. Letter from Richard Gitter, East Coast Director, Broadcast Standards and Practices, ABC, to the New York Chapter of Dignity, Jan. 16, 1975 (courtesy of the National Gay Task Force).
18. National Gay Task Force, "Media Manual to Gay Groups."
19. Interview with Ginny Vida, National Gay Task Force, by Martha Schley, WGBH, Boston, Feb. 3, 1979.
20. Interview with Deiter, April 11, 1979.
21. Women's Media Campaign, Women's Media Project, NOW Legal Defense Fund, Washington, D.C., 1985. Cherie Lewis, *Television License Renewal Challenges by Women's Groups* (unpublished doctoral dissertation, Univ. of Minnesota, 1984).
22. "Presentation, Program Practices," National Gay Task Force Internal Memo, Jan. 17, 1978 (courtesy of the National Gay Task Force).
23. Letter from Lotman, Media Director, National Gay Task Force, to Elton Rule, President, American Broadcasting Companies, Aug. 26, 1975 (courtesy of the National Gay Task Force).
24. This description of the plot of "Flowers of Evil" is based on interviews with Vida and Lotman. Interview with Vida, May 2, 1979. Interview with Herminio Traviesas, Oct. 15, 1978.
25. Interviews with Deiter. See also Richard Lewis, "Putting on a Gay Face," *L.A. Weekly*, April 16–22, 1982.
26. Lewis, "Putting on a Gay Face."
27. Interview with Deiter, July 17, 1979.
28. Lewis, "Putting on a Gay Face."
29. Ibid.
30. Ibid.
31. Montgomery, *Gay Activists and the Networks*.
32. Levine, "How the Gay Lobby Has Changed Television."
33. See Geoffrey Cowan, *See No Evil* (New York: Touchstone, 1978), 262. "Taming a Lusty Show: Censor's Memo Tells How," *Los Angeles Times*, June 27, 1977. "*Soap*'s Premise So Much Suds," *Chicago Sun Times*, Sept. 13, 1977.
34. "Television '77," *Chicago Daily News*, Sept. 12, 1977. "*Soap* a Winner Despite Protest," *Chicago Tribune*, Sept. 15, 1977.
35. Interview with Deiter, July 17, 1979. Letter from Deiter, Gay Media Task Force, to Tom Kersey, Broadcast Standards and Practices, ABC, Los Angeles, May 17, 1977. Although at this point NGTF

had not organized a protest, a small group of gays calling itself the "International Union of Gay Athletes"—which NGTF representatives claimed was "a paper organization of about twelve people"—staged a protest of the program and planned a sit-in at ABC's West Coast offices of broadcast standards. The group acted independently of NGTF, but informed the organization of its plans. ABC met with the International Union of Gay Athletes and invited Deiter to participate in the meeting. Interview with Deiter, Aug. 17, 1979.

36. Morrie Gelman, "Silverman Trying Hard to Keep 'Soap,' " *Daily Variety*, July 18, 1977.
37. Specific changes promised by the network—resembling those suggested earlier by Deiter—were summarized in a memo from GMTF to NGTF: "(1) They are going to have Jodie drop the sex-change operation idea after the first couple of episodes, (2) the character of Jodie is going to be strengthened; he's going to confront his father who treats him like a doormat, and his brother who doesn't want to know he's gay, also, (3) Jodie is going to have a relationship (become lovers with a non-stereotypical football player type gay man)."
38. Advertisement, *Variety*, Sept. 7, 1979.
39. Letter from Ginny Vida, National Gay Task Force, to Fred Silverman, President, ABC Entertainment, Sept. 6, 1977 (courtesy of the National Gay Task Force).
40. Interview with Vida.

Chapter 6. He Who Pays the Piper

1. Richard Levinson and William Link, *Stay Tuned: An Inside Look at the Making of Prime-Time Television* (New York: St. Martin's, 1981), 204–8.
2. Geoffrey Cowan, *See No Evil* (New York: Touchstone/Simon & Schuster, 1978), 74. See also: Willard D. Rowland, Jr., *The Politics of Television Violence* (Beverly Hills: Sage, 1983).
3. Cowan, *See No Evil*, 67–68. Cowan's book provides a detailed account of the controversy surrounding this movie, as well as the events leading up to and following the institution of the Family Hour. The book also includes a discussion of the pressure efforts to reduce violence.
4. Ibid., 67–69.
5. Ibid., 113. As Cowan explains, though the Family Hour was offi-

cially declared invalid as part of the NAB code, it was adopted independently by each of the three networks, where it continued to be enforced. John Revett, "Angry Viewers Flood FCC with Letters Ripping Programs, Citing Advertisers," *Advertising Age*, April 21, 1975.

6. Rowland, *The Politics of Television Violence*, 180. John Revett, "Angry Viewers Flood FCC."
7. John Revett, "Group Pressures FCC to Curb TV Violence, Sex," *Advertising Age*, Sept. 22, 1975.
8. There were still occasional single sponsored programs, but these were exceptions to the common industry practice of spot advertising.
9. Letter from American Motors Advertising and Research to Church of the Brethren, March 14, 1977 (courtesy of the Interfaith Center on Corporate Responsibility).
10. Maurine Christopher, "Execs from GF and O & M Urge Moves Against TV Sex/Violence," *Advertising Age*, Nov. 24, 1975.
11. Howard Eaton, Senior Vice President—Broadcasting, Ogilvy & Mather, General Food Agency, quoted in Maurine Christopher, "Execs from GF and O & M Urge Moves Against TV Sex/Violence," *Advertising Age*, Nov. 24, 1975.
12. Philip H. Dougherty, "Action on TV Violence Urged," *New York Times*, Nov. 21, 1975. Christopher, "Execs from GF and O & M Urge Moves."
13. Christopher, "Execs from GF and O & M Urge Moves." Dougherty, "Action on TV Violence Urged."
14. Telephone interview with Nicholas Johnson, Sept. 10, 1985. Ronald G. Slaby, Gary P. Quarforth, and Gene A. McConnadrie, "Television Violence and Its Sponsors," *Journal of Communication* (Winter 1976).
15. Interview with Johnson. "Violence Count Finds Decline in Family Time, Nowhere Else," *Broadcasting*, April 5, 1976.
16. Interview with Johnson.
17. Les Brown, "Study Assails Sponsors on TV Violence," *New York Times*, July 30, 1976. "NCCB Ties Together Advertisers and Violent Programs," *Broadcasting*, Aug. 2, 1976.
18. "NCCB Ties Together Advertisers and Violent Programs."
19. "No Let-up on Some Fronts in Attacks on TV Violence," *Broadcasting*, Dec. 22, 1975.
20. *The National PTA Handbook*, p. 124. "PTA Declares War on Television Violence," *Los Angeles Times*, Sept. 25, 1976.

21. Richard Cheverton, "PTA Puts Networks 'On Trial' for Airing Crimes of Violence," *Chicago Tribune*, Oct. 6, 1976.
22. "Judge Indicts Video Violence," *Los Angeles Times*, Dec. 3, 1976. "Danish Tells PTA That Boycotts Could Backfire," *Broadcasting*, Jan. 24, 1977. "Television Violence Assailed at PTA Hearings in Chicago," *Los Angeles Times*, Jan. 26, 1977.
23. Phillip H. Dougherty, "Thompson Scores TV Violence," *New York Times*, June 9, 1976.
24. Dougherty, "Action on TV Violence Urged." "JWT Scores Violence in Media at AAF Meet," *Advertising Age*, June 14, 1976.
25. "Procter & Gamble—TV Violence," *Policy Statement*, 1977. "Recruits Join PTA Campaign to Pacify TV," *Broadcasting*, Jan. 31, 1977.
26. Letter from the Legal Department at Eastman Kodak Company, January 19, 1977, to Rev. Michael H. Crosby (courtesy of Interfaith Council on Corporate Responsibility).
27. Ibid.
28. Todd Gitlin, *Inside Prime Time* (New York: Pantheon, 1983), 256–57. Interview with Aaron Cohen, Grey Advertising, New York, June 24, 1983. As one advertiser explained the process: "We feel many programs are generally acceptable but may sometimes have specific episodes which may not be in keeping with our objectives. [In these programs] we are represented only after particularly careful pre-screening of the specific episodes on which our participation is proposed and consultation among our screening service, our advertising agency, and ourselves if there is a question. During the Fall Quarter alone there were at least five instances when we specifically instructed the networks to take us off such episodes after pre-screening." Letter from John R. Preston, Vice President—Marketing, Campbell Soup Company, to Stewart M. Hoover, Church of the Brethren, March 7, 1977 (courtesy of Interfaith Council on Corporate Responsibility).
29. "Networks Think It's for Real as Advertisers Scramble for Antiviolence Bandwagon," *Broadcasting*, Feb. 14, 1977.
30. Interview with Bettye Hoffman, NBC, New York, June 23, 1983. Rowland, *The Politics of Television Violence*, 218–22. Interview with Horst Stipp, NBC Social Research Department, New York, June 23, 1983. "More Violence Than Ever, Says Gerbner's Latest," *Broadcasting*, Feb. 28, 1977. "Violence: What People Think, What TV Is Doing About It," *Broadcasting*, Feb. 7, 1977. "A Blizzard of Paper on Violence," *Broadcasting*, May 16, 1977.
31. "AMA Pushes Anti-Violence," *Advertising Age*, April 4, 1977. "United Church Takes on Mission Agaist Sex and Violence on TV,"

Broadcasting, April 11, 1977. Bryce Nelson, "PTA Warns TV to Cut Violence," *Los Angeles Times,* April 16, 1977.

32. Lee Margulies, "Monitor of Violence," *Los Angeles Times,* Aug. 15, 1977.
33. Cowan, *See No Evil,* 250. "As We See It," *TV Guide,* Aug. 27, 1977. "NCCB: Violence Declined in Summer; Burger King's Turnabout Praised," *Broadcasting,* Sept. 12, 1977.
34. Interview with Jean Dye, National Parents-Teachers Association, Chicago, April 18, 1983. Cowan, *See No Evil,* 266–67.
35. Cowan, *See No Evil,* 257–66.
36. "NCCB had altered its operational definition of violence in response to the criticism it had received. It now used a "murder and mayhem" concept, defined as "the realistic portrayal of a gunfight, gun threat, gun shooting at a person, beating threat, beating, strangling, manhandling, fist fight, inflicting wounds, stabbing, attempted drowning, attempted suicide, killing, kidnapping or suicide." "NCCB Study Details Drop in TV Violence," *Broadcasting,* Feb. 6, 1978. Interview with Johnson.
37. Ron Aldridge, "New PTA Homework: Learning to Live with TV," *Chicago Tribune,* Oct. 7, 1982. Interview with Grace Foster, Parent-Teachers Association, Los Angeles, June 15, 1983. PTA Newsletter, March 1983; interview with Warren Ashley, NBC, Nov. 15, 1984. Margulies, "PTA Takes New Approach to TV," *Los Angeles Times,* Jan 29, 1982. Interview with Dye.
38. Interview with Ashley. Interview with Foster.
39. Interview with Dye. *PTA Newsletter,* April 18, 1983.
40. See D. Pearl, L. Bonthilet, and J. Lazar, eds., *Television and Human Behavior: Ten Years of Scientific Progress and Implications for the Eighties* (Rockville, MD: National Institute of Mental Health, 1982).
41. Telephone interview wtih Sally Steenland, National Commission on Working Women (formerly with NCCB), Aug. 6, 1987. Interview with Brian Malloy, National Coalition on Television Violence, Washington, D.C., June 28, 1983.
42. "TV Sponsors are Studying the Issue of Violence," *Broadcasting,* March 28, 1977.
43. Interview with Robert Kalaski, International Association of Machinists, Washington, D.C., July 1, 1983. "Nationwide Television Monitoring Will Cheer Images of Workers, Unions," *The Machinist* (Oct. 1979).

Chapter 7. Battle over *Beulah Land*

1. Jack Slater, "Confrontation on Minorities on TV," *Los Angeles Times*, Feb. 6, 1980.
2. Howard Rosenberg, "Protests Erupt over *Beulah Land*," *Los Angeles Times*, March 3, 1980.
3. *Broadcasting*, Sept. 1, 1980. David Cuthbert, "*Beulah* Producers Reply." *New Orleans Times-Picayune*, March 3, 1980.
4. Rosenberg, "Protests Erupt over *Beulah Land*."
5. Though petitions were still being filed by citizens' groups in the late seventies, the FCC was slow to act on them. Most were ultimately settled in the broadcaster's favor. National Black Media Coalition chairman Pluria Marshall told a gathering of broadcasters in 1978 that "NBMC and other minority groups were becoming impatient with repeated denials of petitions to deny and might increasingly turn to economic boycotts as a method of communicating their complaints to broadcasters." Barry Cole and Mal Oettinger, *Reluctant Regulators* (Reading: MA: Addison-Wesley, 1978), 304.
6. Standards and practices executives at all three networks told me in the late seventies that they did not have regular dealings with black activist groups. As for the IMAGE award, Gerber told the press the NAACP gave him the IMAGE award for an NBC development project called "The Neighborhood" which dealt with a black family moving into a white neighborhood. "Gerber Counters Planned 'Beulah' Protest," *Daily Variety*, March 3, 1980.
7. Todd Gitlin, *Inside Prime Time* (New York: Pantheon, 1983), 180.
8. Interview with Charlie Hauck, Los Angeles, July 11, 1985. Interview with Norman Lear, Nov. 13, 1985.
9. Interview with Lear. Interview with Virginia Carter, Los Angeles, Nov. 27, 1984.
10. J. Fred MacDonald, *Blacks and White TV: Afro-Americans in Television since 1948* (Chicago: Nelson Hall, 1983), 181. "Is This the Price We Pay for Media 'Success'?" *Los Angeles Sentinel*, Feb. 28, 1980.
11. See Thomas Cripps, "*Amos 'n' Andy* and the Debate over Racial Integration," in John E. O'Connor, ed., *American History/American Television* (New York: Frederick Ungar, 1983). Don Shirley, "Storm over *Beulah Land*," *Panorama* (May 1980).
12. David Cuthbert, "TV *Beulah Land* Worse Than *Mandingo?*," *New Orleans Times-Picayune*. Feb. 18, 1980.

13. David Cuthbert, *"Beulah Land* Brouhaha," *New Orleans Times-Picayune,* March 2, 1980. Cuthbert, "TV *Beulah Land* Worse Than *Mandingo?"*
14. Cuthbert, *"Beulah Land* Brouhaha."
15. Howard Rosenberg, "Protests Erupt over *Beulah Land," Los Angeles Times,* March 3, 1980.
16. Robert Price and Saundra Sharp, "A Position Paper Against the Airing of *Beulah Land,"* Feb. 1980 (courtesy of Robert Price).
17. Cuthbert, "TV *Beulah Land* Worse Than *Mandingo?"*
18. Cuthbert, *"Beulah Land* Brouhaha." Interview with David Gerber, Culver City, Jan. 11, 1985. Jack Slater, *"Beulah Land:* A Case Study of a Protest," *Los Angeles Times,* Oct. 6, 1980. Gerber took out his own ad in the trades in which he explained, ". . . I was willing to attend a requested meeting with them [coalition members] to discuss their concern. But before the scheduled meeting, they took out trade ads condemning the picture knowing full well that the meeting was to be held within a few days." See "No Apologies for Beulah Land," *The Hollywood Reporter,* Oct. 7, 1980.
19. Rosenberg, "Protests Erupt over *Beulah Land."* Cuthbert, *"Beulah Land* Brouhaha."
20. Interview with Robert Price, Los Angeles, Aug. 25, 1980.
21. Ibid.
22. Price and Sharp, "Position Paper Against the Airing of *Beulah Land."*
23. Telephone interview with David Cuthbert, Jan. 16, 1984. Cuthbert, "TV *Beulah Land* Worse Than *Mandingo?"*
24. Cuthbert, "TV *Beulah Land* Worse Than *Mandingo?"*
25. David Cuthbert, "There's More Trouble in *Beulah Land," New Orleans Times-Picayune,* Feb. 22, 1980. Interview with Price.
26. Marilyn Beck, "Blacks Seek to Halt NBC's *Beulah Land," Richmond Times-Dispatch,* March 3, 1980. Beck, "Gerber Counters Planned Beulah Protest," *Daily Variety,* March 3, 1980. David Cuthbert, "Beulah Producers Reply," *New Orleans Times-Picayune,* March 3, 1980. Beck, "Land Sakes, NAACP Is Up in Arms," *Valley News,* March 3, 1980. Beck, "Blacks Seek to Halt NBC's *Beulah Land."*
27. Beck, "Blacks Seek to Halt NBC's *Beulah Land."*
28. Interview with Price.
29. Ibid.
30. *Daily Variety,* March 7, 1980.
31. Cuthbert, *"Beulah Land* Brouhaha."
32. *Los Angeles Sentinel,* March 13, 1980.
33. Interview with Jay Rodriguez, NBC, Burbank, Nov. 30, 1984.

34. Interview with Price.
35. Rick DuBrow, "*Beulah Land* Debate at NBC," *Los Angeles Herald Examiner*, March 12, 1980.
36. Letter from Jay Rodriguez to Robert Price, March 14, 1980 (courtesy of the Black Anti-Defamation Coalition).
37. DuBrow, "*Beulah Land* Debate at NBC."
38. Letter from Robert Price to Jay Rodriguez, March 17, 1980. (courtesy of the Black Anti-Demfamation Coalition).
39. Howard Rosenberg, "More on *Beulah Land*," *Los Angeles Times*, March 10, 1980. Gerald Jordan, "An Imbalance in TV Roles for Blacks," *Kansas City Star*, March 16–22, 1980. Lee Paige, "Urge Blacks to Boycott NBC's *Beulah Land*," *Chicago Metro News*, March 29, 1980.
40. Letter from Augustus Hawkins, Calfornia Congressman, 29th District, to NBC and David Gerber, March 25, 1980 (courtesy of Black Anti-Defamation Coalition).
41. Letter from Lawrence White, President, Columbia Pictures Television, to Augustus Hawkins, March 28, 1980 (courtesy of Black Anti-Defamation Coalition).
42. Letter from Brandon Tartikoff to Augustus Hawkins, April 4, 1980.
43. *Daily Variety*, March 31, 1980.
44. Rick DuBrow, "*Beulah* Postponed; Producer 'Furious,' " *Los Angeles Herald Examiner*, April 5, 1980.
45. Ibid. Viki Pipkin, "Groups Protest Special," *Inglewood/Hawthorne Wave*, April 24, 1980.
46. Dave Kaufman, "Gerber Blasts *Beulah* Postponement," *Daily Variety*, April 7, 1980.
47. A historian had also been hired to work as a consultant on location in Natchez, Mississippi, but he did not continue his involvement with the project after the shooting. Don Shirley, "Storm over *Beulah Land*," *Panorama*, May 1980.
48. Interview with Tilden Edelstein, Rutgers University, New Brunswick, March 28, 1983.
49. Ibid.
50. Eunice Field, "*Beulah* Author Demands His Name Taken Off Credits," *The Hollywood Reporter*, April 28, 1980. Jack Slater, "*Beulah Land:* A Case Study of a Protest," *Los Angeles Times*, Oct. 6, 1980.
51. Field, "*Beulah* Author Demands His Name Taken Off Credits."
52. Howard Rosenberg, "More Trouble in *Beulah Land*," *Los Angeles Times*, May 5, 1980. DuBrow, "*Beulah* Postponed."
53. Interview with Price.

54. NAACP representative at a closed industry meeting. Bonnie Allen, "*Beulah Land*, in Living Colored," *Essence* (June 1980).
55. *Hollywood Reporter*, May 19, 1980.
56. Letter from Jay Rodriguez, NBC to Robert Price, May 6, 1980 (courtesy of Black Anti-Defamation Coalition).
57. Juneteenth Day is a celebrated day in black history in memory of this date in 1865 when slaves were freed in Texas.
58. Sondra Lowell, "Hard Questions for Tartikoff," *Los Angeles Times*, June 23, 1980.
59. Ibid.
60. "Coalition Adamant on *Beulah* Show," *Hollywood Reporter*, Sept. 10, 1980.
61. Alan L. Gansberg, "NBC OK's Final Version of *Beulah*; 1 Cut Sans Gerber," *Hollywood Reporter*, Aug. 14, 1980.
62. Rick DuBrow, "Producer Alters *Beulah Land*, but Denies Pressure," *Los Angeles Herald Examiner*, Aug. 31, 1980.
63. Interview with Price.
64. "Coalition Adamant on *Beulah* Show."
65. DuBrow, "Producer Alters *Beulah Land*."
66. Robert Price and Saundra Sharp, "Second Position Paper Against the Airing of *Beulah Land*," Sept. 15, 1980 (courtesy of Robert Price).
67. Jeff Greenfield, *Sunday Morning*, CBS network, aired Aug. 28, 1980. DuBrow, "Producer Alters *Beulah Land*."
68. Arthur Unger, "A Bargain Basement *Gone With the Wind*," *Christian Science Monitor*, Oct. 6, 1980.
69. Tom Shales, "Stereotypes, Sin, Southern-Fried Trash in NBC's Degrading Miniseries," *Washington Post*, Oct. 7, 1980.
70. Harry F. Waters, "Look Aghast, *Beulah Land*," *Newsweek*, Oct. 13, 1980.
71. "*Beulah Land* Will Encore on NBC," *Los Angeles Times*, Sept. 10, 1983.
72. Per interview with Alfred Schneider, Vice President, Broadcast Standards and Practices, ABC Network, Los Angeles, May 5, 1983.
73. In-house NBC document (courtesy of National Broadcasting Company).

Chapter 8. Cleaning Up TV

1. "NFD's Don Wildmon: The Medium Is the Mission," *Broadcasting*, Feb. 9, 1981. Hodding Carter, "Eye of the Beholder" transcript, *Inside Story*, May 1981.

2. "Pastor Receives Many Inquiries on TV Boycott," *Los Angeles Times,* Feb. 26, 1977. Harry F. Waters, "Does Incest Belong on TV?," *Newsweek,* Oct. 8, 1979. "Where Do We Draw the Line?," *Los Angeles Times,* Oct. 12, 1979. As Todd Gitlin explains: "When ABC, in 1977, scheduled *Soap,* a farcical send-up of sexually indulgent soap operas, it got 32,000 letters of protest, many of them mobilized by Wildmon. When ABC announced it was making a 1980 TV movie from Jarilyn French's feminist novel *The Women's Room,* Wildmon denounced it, sight unseen, as antifamily and letters flooded in again. . . . [The] campaign was sufficiently potent . . . to intimidate advertisers. According to Brandon Stoddard, ABC lost ten out of fourteen minutes of *Women's Room* spot ads before the air date. . . ." Todd Gitlin, *Inside Prime Time,* (New York: Pantheon, 1983), 250.
3. John Weisman, "He's Counting Every Jiggle and Cussword," *TV Guide,* March 17, 1979.
4. Christopher Sterling and John M. Kittross, *Stay Tuned: A Concise History of American Broadcasting* (Belmont, CA: Wadsworth, 1978), 413–14.
5. Thomas Love, "Moral Majority Preparing to Take on TV Networks," *Washington Star,* Nov. 12, 1980. Bill Abrams, "New Target: 'Immoral' TV," *Wall Street Journal,* Nov. 6, 1980.
6. Gitlin, *Inside Prime Time,* 251. David Crook, "Coalition to Monitor TV Programming," *Los Angeles Times,* Feb. 2, 1981.
7. William Proctor, "Television and Decency," *Pentecostal Evangel,* March 22, 1981.
8. Ibid.
9. Gitlin, *Inside Prime Time,* 247–63.
10. Proctor, "Television and Decency." Gitlin, *Inside Prime Time,* 251–52.
11. Carter, "Eye of the Beholder."
12. Kenneth R. Clark, "NBC: The Battle of the Boycotters," *Los Angeles Times,* March 30, 1981.
13. Gitlin, *Inside Prime Time,* 247–63.
14. Clarke Taylor, "CBS to Resist Monitoring Group," *Los Angeles Times,* April 10, 1981. Thomas Love, "Moral Majority Preparing to Take on TV Networks," *Washington Star,* Nov. 12, 1980.
15. Jan R. Van Meter, "Boycotting TV Advertisers: A Message for the Medium," *New York Times,* Oct. 25, 1981.
16. David Crook, "Coalition to Monitor TV Programming," *Los Angeles Times,* Feb. 2, 1981. Proctor, "Television and Decency," 10–11.

17. George F. Will, "Boycotts over TV Sleazery Make Sense, Not Censorship," *Los Angeles Times*, Feb. 9, 1981.
18. Lee Margulies, "The Proliferation of Pressure Groups in Prime Time Symposium," *Emmy Magazine* (Summer 1981). There is some question about whether sponsor boycotts are considered "secondary boycotts." According to David Robb: "The much-misunderstood law of secondary boycotts is strictly a creature of labor law, and simply put, states that it is unlawful to boycott a business (i.e., a grocery store selling grapes, for instance) that is not directly involved in, but, is caught in the middle of a labor dispute (i.e., striking farm workers and grape growers), and in all probability does not apply to such political actions as the DGA's [Director's Guild of America] boycott of states that haven't ratified the ERA." Robb also suggests that the coalition's shift from the term "boycott" to "selective buying campaign" may have stemmed from fear of lawsuits because of the fuzzy legal nature of secondary boycotts. Scott Robb, "Boycott Blues: Shifting Sands of Discontent Sweep Hollywood," *Hollywood Reporter*, Dec. 28, 1981.
19. Gitlin, *Inside Prime Time*, 254.
20. "ABC Study Reveals Product Boycott a Minimal Influence," *Daily Variety*, May 6, 1981. Tony Schwartz, "Studies by 2 Networks Dispute Moral Majority," *New York Times*, June 19, 1981.
21. Schwartz, "Studies by 2 Networks."
22. "TV Networks Issue Studies on Boycott That Say It'll Fizzle," *Wall Street Journal*, June 19, 1981.
23. CBS was the only network to refuse to participate in the meeting. Instead, a televised debate was arranged between Gene Mater and Wildmon a few weeks later. Lee Margulies, "CBS, Coalition Harden Their Positions," *Los Angeles Times*, June 13, 1981. Howard Rosenberg, "Pressure Groups and TV—The Shadow over Ojai," *Los Angeles Times*, May 13, 1981.
24. Howard Rosenberg, "Networks, Advertisers Respond to 'Different Kinds of Falwells,' " *Los Angeles Times*, May 12, 1981. Rosenberg, "Pressure Groups and TV."
25. Rosenberg, "Networks, Advertisers Respond." Margulies, "CBS, Coalition Harden Their Positions."
26. N. R. Kleinfield, "TV Moral Monitors to Press Sponsors," *New York Times*, May 27, 1981.
27. These figures were for the total number of individual episodes, not entire series. Tony Schwartz, "TV Sponsor's Guidelines Called 'Conservative,' " *New York Times*, June 18, 1981. Tony Schwartz,

"50 TV Shows Rejected by Procter & Gamble," *New York Times*, June 17, 1981.

28. Nancy Yoshihara, "Was there a Deal to Nip Boycott?," *Los Angeles Times*, July 7, 1981. Schwartz, "50 TV Shows Rejected."
29. Schwartz, "50 TV Shows Rejected."
30. Herbert Mitgang, "Lear TV Ads to Oppose the Moral Majority," *New York Times*, June 25, 1981. Leslie Ward, "Rallying Round the Flag: Norman Lear and the American Way, *American Film* (Oct. 1981). A number of other advocacy groups took public positions against the coalition's campaign. "It is censorship in its most obvious form," charged Peggy Charren, president of Action for Children's Television, "and I worry more about that than I do about television." "Conservative Group to Monitor TV," *New York Times*, Feb. 3, 1981.
31. Joseph B. Treaster, "TV Advertisers Meet Coalition in Effort to Avert Boycott," *New York Times*, June 26, 1981.
32. Jerry Falwell, Moral Majority contribution letter, June 26, 1981.
33. Tony Schwartz, "Group 'Leaning' to Postponing Boycott of TV Advertisers," *New York Times*, June 27, 1981.
34. Yoshihara, "Was There a Deal to Nip Boycott?"
35. Todd Gitlin, "The New Crusades: How the Fundamentalists Tied up the Networks," *American Film* (Oct. 1981).
36. Falwell contribution letter. Gitlin, "The New Crusades," 62.
37. Ad Boycott Threat Renewed," *Los Angeles Times*, Aug. 13, 1981. Van Meter, "Boycotting TV Advertisers."
38. Jan R. Van Meter, "TV Must Brace for a Boycott," *Los Angeles Times*, Nov. 20, 1981.
39. Tony Schwartz, "A Boycott of TV Advertisers Is Again Threatened," *New York Times*, Jan. 28, 1982. Charles Austin, "Television Sponsor Boycott," *New York Times*, Feb. 14, 1982. "Wildmon to Make Another Stab at Boycott," *Broadcasting*, Feb. 1, 1982.
40. Theresa McMasters, "Wildmon's Minions Will Boycott NBC, RCA Products, Services," *Hollywood Reporter*, March 5, 1982.
41. "TV Boycott Coalition Apologizes to Company," *New York Times*, March 6, 1982.
42. Howard Rosenberg, "Controversy over *'Sister,'* " *Los Angeles Times*, June 2, 1982.
43. Ibid.
44. Interview with Aaron Cohen, Grey Advertising, New York, June 24, 1983.
45. Rosenberg, "Controversy over *'Sister.'* "

46. " 'Sister, Sister' and Rev. Wildmon," *Los Angeles Times*, June 17, 1982.
47. Philip Sheno, "The Boycott Against RCA-NBC," *New York Times*, Nov. 21, 1982. "RCA's Profits down 6.5% in Quarter," *Los Angeles Times*. Peter Boyer, "NBC's Bullish Reversal Explained by RCA Chief," *Los Angeles Times*, May 4, 1983.

Chapter 9. The Hollywood Lobbyists

1. Michele Willens, "Norman Fleishman, Hollywood's Liberal Pied Piper," *California Journal*, Aug. 1983. Microsecond pamphlet (courtesy of Norman Fleishman).
2. For an excellent description of the Hollywood entertainment industry subculture, see Todd Gitlin, *Inside Prime Time* (New York: Pantheon, 1983), 115–42.
3. Telephone interview with David Poindexter, Aug. 28, 1985.
4. Barbara Isenberg, "Sex Education via Entertainment," *Los Angeles Times*, Jan. 24, 1980. See also Michele Willens, "Norman Fleishman, Hollywood's Pied Piper," *California Journal* (Aug. 1983).
5. Interview with Norman Fleishman, Los Angeles, Nov. 11, 1984.
6. Ibid.
7. Les Brown, "Wakefield Quits 'James' in TV Dispute," *New York Times*, Jan. 12, 1978. Internal NBC broadcast standards document (courtesy of NBC).
8. Letter from Jerome H. Stanley to Carol Dilfer, Feb. 7, 1978 (courtesy of NBC). Interview with Fleishman. Richard Hack, "Dan Wakefield Quits 'James' over Tiff with Censors," *Hollywood Reporter*, Jan. 12, 1978. Brown, "Wakefield Quits 'James.' "
9. Interview with Poindexter. Interview with Fleishman. A separate organization also lobbied entertainment television on similar issues as those of the Population Institute. The Project on Human Sexual Development was created in 1975 with a grant from John D. Rockefeller III to "expand the public's understanding of sexuality and its role in every facet of life." The organization formed a "TV advisory committee." In March 1977, the project, headed by Elizabeth J. Roberts, held a three-day seminar in Ojai, California, with network programming and standards and practices executives, producers, writers, and studio heads to educate and sensitize industry representatives to sexuality issues. See Lee Margulies, "Seminar Held on TV Sexuality," *Los Angeles Times*, March 21, 1977.

Interview with Marcy Kelly, Los Angeles, May 10, 1985. Interview with Susan Newman, Los Angeles, Oct. 16, 1986.

10. Interview with Kelly. In 1986, CPO held a one day conference—in cooperation with the Academy of Television Arts and Sciences with support from the Carnegie Foundation—on "Television and Teen Sexual Behavior."
11. The National Commission on Working Women also released periodic monitoring reports on the representation of working women in entertainment television. Zan Dubin, "Asians Cite Those Who Do the Balancing Act," *Los Angeles Times*, March 16, 1985.
12. Joseph Finnegan, "Finnegan's File," *Emmy Magazine* (Sept./Oct. 1986), 136. Interview with Judy Greening, Human Family Education and Cultural Institute, Los Angeles, Aug. 20, 1986.
13. John Horn, "Portrayal of Gays Honored," *Los Angeles Times*, Sept. 19, 1984. Lee Grant, "Gays Give Credit—Where Credit is Due," *Los Angeles Times*, Oct. 6, 1982.
14. Interview with Chris Uszler, Alliance for Gay and Lesbian Artists, Santa Monica, July 12, 1985. Horn, "Portrayal of Gays Honored." AGLA has protested occasionally, but never with the kind of large-scale militancy as the National Gay Task Force. In 1984, when the producers of ABC's *Dynasty* decided to have the gay character, Steven Carrington, marry a woman, a number of gays spoke out against it. AGLA's Chris Uszler's statement to the press over the controversy expressed some concerns, but also indicated tolerance and patience toward the show's creators. Having given the series an award the year before, AGLA was "disappointed by the changes [in Steven]," Uszler exlained. "But it is not unexpected. . . . All along, they've had the character gay/not gay for plot and ratings purpose, and we'll probably see him swing back the other way." John M. Wilson, "Is Steven Gay? Yes, 'Dynasty' Creators Say," *Los Angeles Times*, Feb. 26, 1984.
15. Interview with Tari Susan Hartman, Los Angeles, Sept. 11, 1985. The media office also formed special committees within the industry guilds around the issue of disability, focusing on employment and representation. Similar committees had been set up in the Screen Actors Guild and the Writers Guild on behalf of women and minorities.
16. Interview with Hartman.
17. This shift in concerns is evidenced by AGLA's statements to the press during the organization's awards ceremonies. At the 1983 awards, AGLA singled out several programs for their "happens

to be gay" characters. By 1987, the emphasis was on the fact that "there's more to gay life than AIDS." Michael Lonoon, "Depiction of Gays Honored by Group," *Los Angeles Times*, Sept. 21, 1983. Deborah Caulfield, "Gay Alliance Awards—E for Effort, Emotion," *Los Angeles Times*, March 23, 1987.

18. Interview with Marcy Kelly, May 5, 1987. By 1987 the federal government had begun to urge the TV industry to address the AIDS issue more directly in its programming. At a meeting of the Caucus of Producers, Writers, and Directors in the fall of 1987, for example, Surgeon General C. Everett Koop strongly encouraged the creative community to incorporate references to "safe sex" in its programs.
19. Warren Breed and James R. De Foe thoroughly described their strategy and its results in their "Effecting Media Change: Cooperative Consultation." *Journal of Communication* (Spring 1982).
20. Ibid. Telephone interview with Warren Breed and James De Foe, July 12, 1985.
21. A special "Media-Alcohol Newsletter" was prepared and distributed on a regular basis to industry professionals; the team was careful not to overstate its position. "We do not assume," they explained, "that when the viewer sees Hawkeye and Col. Potter in M*A*S*H sip a martini, the viewer rushes to get a drink. We do assume that as a daily part of the environment the media play a role in how people form notions about approved and disapproved behavior. As an illustration, the social learning and modeling theories hold that if young people see television heroes and heroines drinking and smoking, they are more likely to perceive the cultural environment as favorable to these behaviors." Warren Breed and James De Foe, "Drinking and Smoking on Television, 1950–1982," *Journal of Public Health Policy* (June 1984). Breed and De Foe, "Effecting Media Change." See also L. Wallack, W. Breed, and J. Cruz, "Alcoholism on Prime Time Television," *Journal of Studies on Alcohol* (Jan. 1987).
22. Breed and De Foe, "Effecting Media Change."
23. Open letter from Jim De Foe to media professionals (courtesy of A.I.M.S.).
24. Breed and De Foe, "Effecting Media Change." "Four Stages in Cooperative Consultation," undated document (courtesy of A.I.M.S.).
25. Breed and De Foe, "Effecting Media Change."
26. Ibid.

27. Ibid.
28. Interview with Larry Stewart, Entertainment Industries Council, Los Angeles, Aug. 1, 1985.
29. Christopher Sterling and John M. Kittross, *Stay Tuned: A Concise History of American Broadcasting* (Belmont, CA: Wadsworth, 1978), 395–96. "New Threat to Alcohol Advertising," *Broadcasting*, Nov. 28, 1983. Interview with Stewart. "Broadcasters Take Offensive with Drug, Alcohol PSA Campaign," *Broadcasting*, Oct. 29, 1984.
30. Testimony presented on behalf of the Caucus of Producers, Writers, and Directors by Larry Stewart—Permanent Subcommittee on Investigations, U.S. Senate, March 20, 1985.
31. National Council on Families and Television, "Information Service Bulletin," Feb. 1987. Tom Girard, "AFL-CIO Sharpens TV Profile," *Daily Variety*, April 25, 1984. Interview with Kathy Garmezy and Gina Blumenfeld, Los Angeles, Oct. 12, 1985.
32. "A Proposal for the Media Project," Center for Renewable Resources (courtesy of Tyrone Braswell).
33. Interview with Tyrone Braswell. According to Braswell, solar dealers told him that the *All My Children* solar house had influenced buyers to purchase solar units for their own homes. Burt Solomon, "The Selling of Solar Shifts to Hollywood," internal newsletter, *Solar Lobby* (courtesy of Tyrone Braswell).
34. Solomon, "The Selling of Solar Shifts to Hollywood."
35. Interview with Braswell.

Chapter 10. Packaging Controversy

1. Interview with Daniel Donehey, National Right to Life Committee, Washington, D.C., March 4, 1986.
2. Rick Du Brow, "TV Networks No Longer Afraid of Courting Controversy," *Los Angeles Herald Examiner*, Jan. 31, 1984.
3. Joanmarie Kalter, "If You Want Adult Themes, Hot Issues, Social Relevance—Then Stay Home!," *TV Guide*, Feb. 9, 1985. "*The Day After*: A Gamble That Paid Off!," Press release, J. Walter Thompson USA, Inc., Dec. 1, 1983.
4. Kalter, "If You Want Adult Themes." "*The Day After*: A Gamble That Paid Off!"
5. Richard Zoglin, "Troubles on the Home Front," *Time*, Jan. 28, 1985. Jay Sharbutt, "Tartikoff Seeks More Adult Themes," *Los Angeles Times*, Jan. 13, 1984.
6. Todd Gitlin, *Inside Prime Time* (New York: Pantheon, 1983), 159.

Richard Levinson and William Link, *Stay Tuned: An Inside Look at the Making of Prime-Time Television* (New York: St. Martin's, 1981), 110–11. Interview with Marianne Brussat, Cultural Information Service, New York City, May 1, 1987.

7. Interview with Chris Uszler, Alliance for Gay and Lesbian Artists, Los Angeles, July 12, 1985. Remarks by Kathy Bonk, National Organization for Women, to a meeting of the Communications Network in Philanthropy, Santa Monica, California, Oct. 30, 1986. John Wilson, "Catching TV With Its Taboos Down," *Los Angeles Times,* Aug. 26, 1984.
8. Jack Curry, "Weighty Issues Creep into Sitcoms," *USA Today,* Feb. 13, 1985. Howard Rosenberg, "That 'Uh-Oh' Feeling, and What to Do," *Los Angeles Times,* Jan. 24, 1985.
9. The involvement of standards and practices in the development of prime-time programs was similar to that of the programming department, though the two units within the network served distinct purposes. While programming oversaw the writing and production of the programs with the goal of ensuring the highest ratings possible, standards and practices was charged with seeing to it that no program element could cause trouble for the network. Sometimes the goals of these two departments put them in conflict with one another, but generally their activities were complementary, and both departments shared an interest in helping the network operation run smoothly and profitably. For a description of the structure of operation of standards and practices departments, see Robert Pekurny, *Broadcast Self-Regulation: A Participant-Observation Study of the National Broadcasting Company's Broadcast Standards Department* (unpublished doctoral dissertation, Univ. of Minnesota, 1977).
10. Melvin S. Heller, M.D., "Broadcast Standards Editing," American Broadcasting Companies, Inc., 1978.
11. NBC's thirty-page "Broadcast Standards for Television" explained its policy regarding stereotyping with the following brief sentence: "Special sensitivity is necessary in presenting material relating to sex, age, race, color, creed, religion or national or ethnic derivation to avoid contributing to damaging or demeaning stereotypes." For years, CBS refused to adopt any written guidelines. One of the network's executives defended this practice by arguing, "If we had a written policy, it would just become a target for special interest groups." The demise of the NAB code resulted from a suit filed by Westinghouse broadcasters to protest the network policies in restricting the number of commercial minutes to

be sold to advertisers. Westinghouse argued that such rules, by artificially restricting the amount of time available for purchase, constituted a form of restraint of trade. The Justice department agreed with the complaint, and the code was dismantled. Peggy Pagano, "Ruling May Bar TV Ad Restrictions," *Los Angeles Times,* July 17, 1982.

12. Interview with Charlie Hauck, Los Angeles, July 11, 1985. Occasionally a dispute might occur and the producer might go over the head of the editor to his or her superior in New York or, infrequently, to the president of the network. But usually any conflicts were worked out much earlier through negotiation. Routines were also established for resolving conflicts. If there were a disagreement over how a scene should be shot, for example, the standards and practices editor might ask the producer to shoot the scene in two different ways, and the final decision as to which version to include would be made when the "rough cut" was screened to the censors. See Pekurny, *Broadcast Self-Regulation,* 318.
13. Interview with Barney Rosenzweig, Los Angeles, Aug. 29, 1986. Interview by Lynne Kirby, UCLA, with Jerilynn Stapleton, National Organization for Women, Los Angeles, Sept. 17, 1985. Interview with Rosenzweig. Other groups participated in the effort to reinstate *Cagney and Lacey,* including newly organized Viewers for Quality Television, which subsequently spearheaded similar letter-writing campaigns to keep shows on the air. Judy Mann, "TV Ratings Rebellion," *Washington Post,* March 8, 1985.
14. Interview with Rosenzweig.
15. Interview with Christopher Davidson, CBS, Los Angeles, Aug. 4, 1986.
16. Interview with Milton Gross, Fairness Division, Federal Communications Commission, March 5, 1986. See also Steven J. Simmons, *The Fairness Doctrine and the Media* (Berkeley: Univ. of California Press, 1978).
17. "Taming a Lusty Show: Censor's Memo Tells How," *Los Angeles Times,* June 27, 1977. Interview with Warren Ashley, NBC Broadcast Standards Department, Los Angeles, Nov. 15, 1984. NBC Internal Broadcast Standards memo (courtesy of NBC).
18. Interview with Ashley. In 1974, gun lobby groups had found out about the ABC TV movie *The Gun* while it was still in production. The film chronicled the biography of a gun as it went from owner to owner, finally ending up in the hands of a small child who accidently killed himself with it. The National Rifle Association launched a massive campaign in protest to *The Gun* and filed a

Fairness Doctrine complaint against ABC, charging that the network had engaged in "unfair, one-sided, inaccurate, anti-handgun broadcasting practices," by broadcasting programs which depicted handguns "in a consistently unfavorable light to the exclusion of the many legitimate uses of privately owned handguns." The complaint asked the Commission to "require ABC to devote equal time to presentations showing the legitimate uses of privately-owned handguns and stating that the vast majority of handguns are never used in crime" and suggested that the National Rifle Association would be "well qualified" to present contrasting viewpoints to ABC's "anti-gun" programming. As in the other Fairness cases, the FCC denied the complaint, but the experience, along with other complaints from the gun lobby, was enough to make the networks particularly careful about the way they treated this issue. See Simmons, *The Fairness Doctrine,* 153.

19. Interview with Davidson, CBS, Los Angeles, Aug. 4, 1986. Interview with Rosenzweig.
20. Interview with Davidson.
21. Interview with Davidson. Interview with Rosenzweig.
22. Ibid.
23. Ibid.
24. Interview with Davidson.
25. Ibid.
26. Interview with Davidson. Interview with Rosenzweig.
27. Interview with Davidson.
28. Howard Rosenberg, "Abortion Issue Almost a Prosaic TV Theme," *Los Angeles Times,* Nov. 8, 1985.
29. Sal Manna, "Sorry, This Show Wasn't Seen," *Herald Examiner,* July 5, 1982.
30. Interview with Rosenzweig. Morgan Gendel, "*Cagney & Lacey* Seeks Help on Abortion Show," *Los Angeles Times,* Oct. 23, 1985. Rosenberg, "Abortion Issue Almost a Prosaic TV Theme." Nanci Hellmich, "Daly Defends 'Cagney' Show on Abortion," *USA Today,* Nov. 6, 1985.
31. Interview with Rosenzweig. Interview with Stapleton.
32. Gendel, "*Cagney & Lacey* Seeks Help."
33. Internal NOW document, fall 1985.
34. Interview with Rosenzweig.
35. Karen Stabiner, "The Pregnant Detective," *New York Times Magazine,* Sept. 22, 1985. Interview with Daniel Donehey, Washington, D.C., March 4, 1986.
36. Interview with Donehey.

37. Ibid. Daniel Donehey, "*Cagney and Lacey:* Propaganda from CBS," *National Right to Life News*, Nov. 21, 1985.
38. "An Episode of 'Cagney' Under Fire on Abortion," *New York Times*, Nov. 6, 1985. Though the protesters also threatened to boycott sponsors, the network told the press it wasn't really worried. "We've been through various types of organized boycotts in the past, as have NBC and ABC," remarked CBS vice president George Schweitzer, "and we don't tend to see that reflected in the ratings." Morgan Gendel, "AIDS and 'An Early Frost': The Whisper Becomes a Shout," *Los Angeles Times*, Nov. 13, 1985.
39. Interview with Rosenzweig.
40. Rosenberg, "Abortion Issue Almost a Prosaic TV Theme."
41. "CBS O & O's picketed over Cagney Seg," *Daily Variety*, Nov. 12, 1985. Gendel, "AIDS and 'An Early Frost.' "

Chapter 11. From Ferment to Feedback

1. "NBC Holds 5th Conference on Television and Society," *Of Special Interest (A Periodical of Fact and Opinion from NBC Corporate Communications)*, March 1986.
2. An argument can also be made that positive roles for women increased as more women were hired or promoted into positions of power within the TV industry. Employment of women no doubt played some role in encouraging more prominent representation of women in programming, but it was the combination of advocacy group pressures, changes in the larger society, and increasing numbers of women in the industry that produced the positive changes in portrayals that did occur. Muriel Cantor discusses a number of these issues in "Women and Diversity," an unpublished report to the Benton Foundation, June 1987. Despite some successes for women in individual programs, researchers in the late 1980s still could see only limited progress for women's roles in the overall prime-time schedule. See Nancy Signorielli, ed., *Role Portrayal and Stereotyping on Television: An Annotated Bibliography of Studies Relating to Women, Minorities, Aging, Sexual Behavior, Health, and Handicaps* (Westport, CT: Greenwood Press, 1985).
3. There were a few successful shows with primarily black casts, which the networks often cited as proof that blacks had indeed made substantial gains in television. Most notable of these was *The Cosby Show*, a highly popular NBC comedy series in the late eighties.

But blacks had still made few inroads into serious drama in prime time.

4. Peter J. Boyer, "Mark Fowler's 5 Years at the F.C.C." *New York Times,* Jan. 19, 1987.
5. The National Citizens Committee for Broadcasting (which changed its name to Telecommunications Research and Action Center, or TRAC) steadily declined in power and influence, due to waning resources and leadership changes. The United Church of Christ's Office of Communication continued its efforts in the broadcasting area, but found its resources sapped when forced to direct much of its energy to the quickly emerging cable industry. Susan Witty, "The Citizen's Movement Takes a Turn," *Channels* (June–July 1981).
6. Penny Pagano, "House OK's Fairness Doctrine," *Los Angeles Times,* June 4, 1987. Deborah Mesce, "Court Asked to Block FCC Decision," *Los Angeles Times,* Aug. 11, 1987.
7. Black Entertainment Television (BET), a black-owned cable network formed in 1980, enjoyed only limited success. Carried on a few cable systems, the small service was available to only 12.5 million homes by the late eighties. In addition to its marginal position in the new media, BET offered primarily entertainment fare, with a minimum of programming devoted to black issues. "BET: Still Small, but Determined," *Broadcasting,* Feb. 8, 1985. See also *Channels '87 Field Guide to the Electronic Environment.*
8. Anita Wallgren, "Video Program Distribution and Cable Television: Current Policy Issues and Recommendations," Report of the National Telecommunications and Information Administration, U.S. Department of Commerce, June 1988. See also Ben Bagdikian, *The Media Monopoly* (Boston: Beacon Press, 1987).
9. Bill Lewis, "Cable Audiences Rise as Broadcast Nets Slide," *Multichannel News,* June 1, 1987. Morgan Gendel, "Takeover Tremors Top Network Agendas," *Los Angeles Times,* May 6, 1985. Paul Richter, "Early Retirement Plan Is Offered to 2,000 at CBS; Layoffs Possible," *Los Angeles Times,* Sept. 4, 1985. Christopher Vaughn, "CBS: Bad Day at Black Rock," *Hollywood Reporter,* July 3, 1986.

   Fairness in Media (FIM) was formed in 1985 by a group of conservative leaders which included South Carolina Republican Senator Jesse Helms. Charging CBS with "liberal bias" in its news, the group vowed to buy the powerful network in order to exert controls over its programming. This takeover attempt was one of the most aggressive moves by conservatives control network television. Though the group focused most of its wrath on the network's news division, many in the Hollywood community were fearful about the consequences such a takeover might have for entertainment programming. The network ultimately fought off the takeover and was subsequently purchased by a "more friendly" businessman, Laurence Tisch. For a description of this episode in CBS's history, see Peter Boyer, *Who Killed CBS?* (New York: Random House, 1987).

10. Interview with Christopher Davidson, CBS Program Practices, Los Angeles, California, Aug. 4, 1986. L. J. Davis, "Looser, Yes, But Still the Deans of Discipline," *Channels* (July–Aug. 1987).
11. As John Wicklein notes, even the public television system is strongly influenced by the forces of commercialism. "Corporate underwriters now select many of PBS's prime-time programs," he explains. "Corporations rarely fund enterprising news programs or investigative documentaries because the ideas expressed might run counter to their commercial interests. . . . Now we have 'enhanced underwriting,' with credits that are often blatant plugs. . . . Many programmers and producers in the system say that the tendency today is to shape content to suit the purpose of potential corporate underwriters." John Wicklein, "The Assault on Public Television," *Columbia Journalism Review* (Jan./Feb. 1986).

Index